To Matthew,
With love
From Pat and Pat.
Christmas 1975.

Creatures
of the World

Peter Bowen

Marshall Cavendish·London & New York

Edited by Linda Doeser

Published by Marshall Cavendish Publications
Limited 58 Old Compton Street, London W1V 5PA
c Marshall Cavendish Publications Limited 1975

First printing 1975

ISBN 0 85685 134 5

Printed in Great Britain by Jarrold and Sons
Limited

Introduction

In this century of advanced technology and computerization it is refreshing to remind ourselves of the freedom and splendour of the natural world. This book is a comprehensive guide to the diverse and prolific animal life of the air, the land and the sea. It studies each group and class of wildlife, their habits and environments, their inter-relationships and their dependence on each other.

Some of the animals you will find in these pages are rare like the panda, or rapidly disappearing like the koala bear. Others are dangerous like the Portuguese man-o'-war, comical like the penguin, magnificent like the peacock or familiar like the house mouse. All of them are fascinating and you will discover many new facts about common and well-known species.

Examples have been included from all parts of the world, from Antarctica to the Amazon, from Greenland to Australia. All the superb photographs of untamed and uncaged animals can teach you more than you could ever discover on a visit to a zoo or marine aquarium.

This beautifully illustrated book is packed with intriguing information for all those who care about the world's wildlife and its chances of survival.

Endpapers *Zebras*
Previous page *Brown Bear*
This page *Pelicans*
Next page *Peafowl*

The following system has been used to indicate which pictures the captions refer to:

◀ left   ▶ right
▲ top   ▼ bottom
◀▲ top left   ▶▲ top right
◀▼ bottom left   ▶▼ bottom right
◀◀ far left or previous page
▶▶ far right or next page
◀◀◀ previous page left
▶▶▶ next page right

Contents

There are many thousands of species of flying animals, but this book will deal only with the flying vertebrates. In the course of the evolution of the vertebrates, animals with backbones, the ability to fly has arisen three times, each independently and in a different class of animals. Most people are aware that there was once a group of flying reptiles, the Pterosaurs, now extinct. They were probably not particularly good flyers, and tended to glide and soar on the thermal currents in the air. There are also the flying mammals, the bats, which will be considered later in this book. Mammals like the so-called flying squirrel are not true flyers, but gliders. However, the most successful group of flyers, and the most specialized of any of the vertebrates, is that of the birds.

The birds evolved from a now-extinct group of reptiles, many millions of years ago. Surprisingly,

Bee-eater *Merops nubicus*
The bee-eaters are commonly thought to be confined to the tropics. These spectacularly-coloured birds are carmine bee-eaters, natives of Africa and southern Asia. There is an Australian bee-eater, known as the rainbow bird, and a European species, widespread in southern Europe during the breeding season and an occasional visitor to Britain. Bee-eaters nest in colonies. Each pair of birds digs a burrow in the soft bank of a river or road. As many as 10 eggs are laid, with both parents incubating the eggs.

perhaps, they did not evolve from the flying reptiles. Modern birds have retained some of the characteristics of their reptilian ancestors. They have scales on their legs and feet, and their toes terminate with claws. Many aspects of their internal anatomy show their close relationship to the reptiles and one of their most prominent characteristics, their feathers, are modifications of reptilian scales. The horny beaks or bills of birds are also derived from scales.

Birds show many interesting adaptations for flight. Perhaps the most obvious is the modification of the front limbs to form the wings. The typical bones of the vertebrate forearm have become shortened and fused, and the smaller bones of the wrist and the hand lengthened and strengthened by tendons to form the large flat surface necessary for the aero-dynamics of flight. Next most prominent are the feathers, which are light-weight and designed to facilitate flight. Birds are the only vertebrates that have feathers, and they serve not only to make flight possible but also to insulate the bird. The feathers overlap, entrapping air and forming a protective layer that reduces the loss of body heat, keeping the body temperature high. A high body temperature, as seen in birds and mammals, together with insulation provided by feathers or fur, has allowed a high degree of activity in these two groups. It enabled them to fan out across the globe, exploiting almost all available living space because they could virtually ignore their surrounding temperatures. Thus birds and mammals are found in the water, on the

ground and in the air on almost every part of every continent, a significant advance over their ancestors the fish, amphibians and reptiles, which were much more restricted, because of their inability to control their body temperature.

The birds' other adaptations to flight are perhaps less immediately obvious. The lungs are modified for a highly-active life. Birds have developed a series of air-sacs which surround the lung. When the bird inspires, the residual air in the lung, air that always remains after expiration, is forced into these sacs. So the air in the lungs is always completely fresh, with a high oxygen content, rather than being polluted by the residual air as happens in other vertebrates with lungs.

Birds have no diaphragm and breathing is accomplished by the muscles of the chest. These are relatively massive in most birds because they are the muscles of flight. They are attached to the prominent keel of the breast bone and when the bird flies, their contraction and relaxation causes the chest cavity to contract and expand similarly. Thus breathing, in flight, is an involuntary activity for the bird, not requiring the expenditure of extra energy.

An examination of the skeleton of a flying bird shows that the bones are both reduced in size and are hollow. This latter characteristic not only provides strong bones but also lighter ones. The bones of the lower part of the backbone and the pelvis have become reduced in size and number, to decrease weight, and have fused, to provide a solid structure to absorb the shock of the

impact on landing. The two halves of the pelvis are not joined together and so the internal organs have shifted backwards, moving the bird's centre of gravity back over the legs to aid balance. The separated pelvis also facilitates the laying of large eggs. The bones of the legs themselves have become fused and strengthened, and the number of toes reduced. The arrangement of muscles and tendons in the legs and feet in the perching birds, mostly the song birds, causes the toes to close involuntarily around the perch when the weight of the body is supported by the legs.

Other internal modifications include the loss of teeth, which are heavy structures. They have been replaced by the bill, a relatively light-weight horny covering for the upper and lower jaws, and the muscular gizzard which has a tough, rough lining and is responsible for masticating and grinding the food. The reproductive organs are also reduced in most female birds. The right ovary and oviduct have atrophied and all eggs are produced by the left ovary.

The brain of birds is also modified to assist them in their way of life. The co-ordination and vision centres are very highly developed. The need for acute, long-distance vision in a fast-moving animal is obvious. By contrast, the centres of smell in the brain have been greatly reduced.

All these modifications for flight are also found in the various flightless birds—the emu, the kiwi, the ostrich, the penguin and others. This is because these birds have evolved from ancestors who could fly. Their inability to fly has been secondarily acquired, to adapt them to exploit a different type of environment. Flight, which allows birds to escape from predators, has been replaced by other characteristics in these birds; speed and strength in the ostrich, diving and swimming in the penguin, and so on. The fact that these birds have survived is proof that they do not necessarily need to be able to fly to be successful. However, several flightless birds, such as the dodo and the elephant bird, have become extinct in recent times due to man's interference and his introduction of new predators into their environments.

The behaviour of birds is also an adaptation to their mode of life, and has been responsible for their survival and success. Most people are aware of the highly ritualized instinctive behaviour peculiar to birds. Patterns of courtship at the beginning of the mating season not only enable individuals of a species to recognize and respond to each other, but also ensure that those who are most active, aggressive and competitive will breed and pass on these characteristics to their offspring. The bright colours and exuberant song of male songbirds in the spring is part of the same phenomenon. By his display, both physical and vocal, he not only attracts a suitable mate, but also warns his fellow males that he has staked out his territory and is prepared to defend it. This territorial behaviour, displayed by many different groups of animals, is most highly developed in the song birds. It reinforces the pair bond between male and female, and ensures

Skylark *Alauda arvensis Skylarks are best known for their song, making them the most popular of the larks. These are young skylark chicks. The nest is built on the ground, in a hollow or depression, and two or three broods of from three to five chicks may be raised in a season. They are birds of the open country, most easily noticed in the spring, during courtship, when the male flies high into the sky, singing with all his might, then plunging to earth, appearing scarcely to clear the ground, and then repeating the process.*

that pairs of birds of the same species will be evenly distributed over all the available breeding area, reducing or eliminating competition for nesting sites and food. As well, there is always a residual population of unpaired birds which do not have a territory. These are usually younger birds, probably in their first breeding season, which are not able to cope with the defence of the older residents. This population serves a purpose as well, however. If some birds should be killed by predators, or if some natural disaster should destroy the territory holders, then there will be another group ready to move in and take over, ensuring the survival of the species.

Behaviour within a species, especially between a territorial male and an intruder, is a fascinating study. By singing and displaying himself, the territorial male warns others to keep away. If an intruder should venture into his territory, or if a neighbour should overstep the boundary, then usually a 'keep out' song will be sufficient to repel the invader. Even if, with this warning, the stranger does not retire and the resident attacks, the actual body contact is either nil or minimal. The territory holder always wins. This means that there is no physical competition and neither party damages the other. The species does not get involved in internal strife and there is no bloodshed or death.

Some species, like the colonial sea birds, are less territorial than the land birds. Their territory may be no bigger than the nest, and perhaps a small area around it. Some birds' territory is mobile in that the male defends the female from other males, so his territory is wherever she happens to be. Most colonial birds have common feeding grounds but all birds appear to have some sort of defended area which can properly be called a territory.

Courtship, mating and the rearing of young are all highly intricate, innate behaviour patterns. They always follow a specific series of events—courtship, nest building, copulation, egg laying and incubation. The rearing and protection of the young is always necessary for a differing period of time. Usually, both parents share the duties of nest building, incubation and the care and feeding of the chicks, and it is only rarely that the female alone takes all the responsibility.

Migration is a phenomenon usually associated only with birds, although some other animals are known to migrate. The origins of migratory behaviour are not clear, although several reasons have been suggested. There is no doubt that it is to a species' advantage to exploit as many resources as it can. Certainly migration does allow birds to move into temperate zones in the warmer months to utilize the greater availability of nesting sites and abundant food supply at a time when great quantities are required. It is impossible to say whether migratory behaviour results from the necessity to exploit these resources, or whether it is a relic of the ice age, when many animals were driven from their usual environments and could only return to them during the relatively warmer months, or moved into northern areas as the ice retreated. But certainly the phenomenon of migration is one of the more spectacular aspects of bird behaviour.

There is much to be said about the almost endless diversity of variation and adaptation of individual species of birds, of their habits and idiosyncrasies.

◄◄ **Sparrowhawk** *Accipiter nisus*
The sparrowhawk is a relatively common bird of prey in British woodlands. They are useful to man in that at least 60 percent of their diet is of small grain-eating birds. Like many other birds of prey, such as the peregrine and the golden eagle, the sparrowhawk has been declining because of the widespread use of DDT and similar insecticides, but with new controls on the use of these chemicals, hopefully these useful and beautiful birds will return in larger numbers.

◄ **Harpy Eagle** *Harpia harpia*
Not much is known about this, the largest of all the eagles, probably because of its distribution through Central and South America. It nests in tall trees in remote areas, and each pair probably only raises one young every two years. The young is fed a diet of monkeys, sloths and other mammals, with occasional parrots and macaws. The young appear to be reluctant hunters, and only learn for themselves when they have been abandoned by their parents.

11

Land birds

It is impossible to classify birds fully according to their habitat, but a few generalizations can be made. Land birds are those which build their nests in dry places, either on the ground or in vegetation, and which feed on land-dwelling animals and plants. They include the most common group of birds, the passerines or song birds, as well as other familiar groups: the birds of prey—owls, eagles and hawks; some of the common game birds—grouse and pheasants, the large ostriches and rails and other flightless birds, and many of the natives of the tropics—parrots and their relatives, the birds of the dry open plains, and so on. Probably the characteristic most common to all these birds, in spite of the wide variety of their individual habits, diets and behaviour patterns, is the possession of a clearly defined

◄ **Kestrel**
The common kestrel is Britain's most common bird of prey, and various species and races of kestrel are found on all continents except Antarctica. Nesting occurs anywhere a flat and sheltered surface is available, although no proper nest is built. Four to five eggs are laid which hatch in May. Recently, kestrels have been nesting even in central London.

◄◄ **Snowy Owl** *Nyctea nyctea*
Although not the largest of the owls, the snowy owl is still a very powerful bird, with the characteristic feathered legs and the large, powerful talons of a bird of prey. Its native breeding ground is the Arctic tundra, and it migrates south to the Shetlands and Orkneys. In times of severe weather or of food shortage it may be driven as far south as England. Unlike other familiar Arctic inhabitants, like the Arctic fox and the ptarmigan, the snowy owl retains its white plumage throughout the year, although the female, who spends a lot of time on the ground in the summer incubating the eggs, is not the brilliant white of the male, and has dark bars which help to camouflage her during this vulnerable period.

and well defended territory. Naturally, the size of the territory may vary considerably, usually in relation to the size of the bird and the density of the population. Territories ensure that the breeding population is widely spread over the suitable areas of a species' environment, and that there is an equitable distribution of nesting sites and food supplies. However, this does not mean that a bird or a breeding pair of birds will spend all its time within the territory. Some species have, in addition to their defended territories, communal areas for feeding, preening, sunbathing and other activities. The communal area may be a small one, such as a feeding station, or a large one, such as an open field or similar piece of land. It is not uncommon to see birds which are known to be highly territorial in small aggregations, existing together without apparent conflict. Thus some birds may have two geographical areas of activity, one the territory, which is defined as

the area which a bird defends against intrusion by members of the same species, and the other the home range, the area over which a bird is active in its day to day life. Obviously, the home range will be the larger of the two, and will contain within it the territory of the individual as well as possibly one or more territories of other birds.

Another characteristic of land birds, although again not exclusive to them, is migratory behaviour. The migration of song birds, although not as spectacular in some ways as that of sea birds, is still an impressive achievement. Migratory birds are, for the most part, those which breed in the temperate zones of the world, and find it necessary to seek out a warmer climate with a good food supply during the winter months. The stimulus which necessitated the development of migratory behaviour is not clear, but there is no doubt that it is an ancient behaviour pattern that has enabled

many species to survive. Migration flights may be as short as a few hundred kilometres or up to several thousand kilometres. Each migratory species shows a specific migratory behaviour. The northward flight begins at very nearly the same time each year, and the return to the wintering grounds is similarly constant. Probably the determining factor is day length, increasing in the spring and decreasing in the autumn, as this is the only constant factor in the environment which could produce such a regular timetable. Local factors, an early onset of winter for example, may cause a slight variation in the timing of migration, but the changes brought about are relatively minor ones. Probably migratory behaviour is controlled by the birds' hormonal system. The lengthening or shortening days register in the brain by means of the amount of light falling on the retina of the eye, and this triggers off a hormonal response in the pituitary gland, the master

gland of the hormonal system. This in turn stimulates those areas of the nervous system controlling migratory behaviour.

Land birds may have generalized flight paths, sometimes several hundred kilometres wide for a single species, although almost certainly individual birds follow the same path from year to year. Other species may follow a narrow and very specific route. Although most migratory flights are at night, the birds almost certainly make use of particular landmarks to assist navigation.

Although many land birds do take part in seasonal migrations, the most spectacular and lengthiest migratory flights are those of the water birds, particularly the marine birds such as the albatross and the Arctic tern. The subject of migration and navigation during flight will be discussed further in the introduction to water birds.

Shetland's Snowy Owl
In recent years, snowy owls have begun to colonize the northernmost Scottish islands. No proper nest is built, and the eggs are laid, and the young reared, on relatively open ground, although the rock-like colouring of the chicks seems to be sufficient camouflage. Eggs are laid at intervals of two days and hatch in the same way, so that the young owls, being different ages, are different sizes.

◄ **Hyacinth Macaw** *Anodorhynchus hyacinthinus*
The macaws are the largest and most brightly-coloured of
the parrot family. The hyacinth macaw is generally
considered the most beautiful, and fetches the highest
prices in pet shops. The macaws are native to Central and
South America and the West Indies. The large and
powerful beak is used for crushing nuts and seeds. Even
very hard nuts, like Brazils, are an easy task for a
macaw, the strong tongue deftly searching out and
extracting the kernel.

◄ ▼ Red and Blue Macaw *Ara Chloroptera*
The red and blue macaw is another of the 18 species of macaw still to be found from Mexico to Brazil. Many species are, unfortunately, now extinct. As their only predator appears to be the harpy eagle, this extinction is probably due to man. Their brilliant colours and flocking habits made the macaws easy prey for unscrupulous collectors in the seventeenth and eighteenth centuries, when macaws were royal status symbols. Their colourful plumage was also prized.

▼ Parrot *Psittacus erithacus*
There are more than 300 members of the parrot family, but only about 100 of them actually bear the name 'parrot'. The best-known are the Amazonian and African species, because they make the best pets. This parrot, the African grey, is reputed to be the best talker of all especially the male bird.
Like the macaws, it has a large and powerful beak for crushing, but all the parrot family eats soft fruit as well, and probably insects, grubs and the like in the wild.

◄ **Vulture** *Gyps coprotheres*
Vultures, the large, well-known scavenging birds of prey, are found in the Americas, Europe, Asia and Africa, but the best-known are the African species, like the Cape vulture pictured here. In flight, soaring on the air currents, they appear graceful and beautiful, but on the ground they are unattractive and, to some, repulsive birds. Their behaviour and appearance does not help their reputation. They feed on large, decaying corpses of mammals, tearing the rotting flesh apart with their large beaks. Their bare heads and almost featherless necks, adapting them to reach into the cavities of their food, do not make them particularly attractive. Their numbers are now declining with increasing agricultural developments and improved hygiene, but their usefulness, in removing carcasses which are a potential source of infection and disease, is frequently underestimated. Their unfortunate appearance tends to prejudice people against them, but they are an extremely useful and necessary part of the ecology. Stories of predation of livestock have not been substantiated.

▶ ▶ **Tailor-bird** *Orthotomus atrogularis*
Tailor-birds are members of the warbler family. They are found in Asia, from India and southern China, through Japan and the Philippines to Australia. They derive their name from the curious way in which they build their nests. It is made of leaves which the bird stitches together to form a pouch which is then lined with soft plant material like tree cotton or kapok. The leaves are joined together by the bird, usually the male, who punches a hole with his beak through the adjoining edges of the leaves and then threads fibres or spiders' webs through it, twisting and spreading the ends of the binding material to form a sort of knob. The tailor-bird breeds throughout the year. Clutches of four eggs are laid, and these are incubated by the female. In the Asian tailor-birds, one species of which is shown here, the male feeds the female while she is incubating, and both parents rear the young. Australian males, however, mate with several females, and the female rears the young alone.

▶▶▶ **Woodpecker** *Dendrocopus minor*
The woodpeckers are found in wooded areas almost all over the world, except for Australia, Malagasy and the oceanic islands. They are usually brightly coloured, can be very large and have a characteristically long sharp-pointed beak. They also have an extremely long tongue, which is useful for helping to catch their prey. Most woodpeckers hunt for their food, insects and larvae, in the crevices of the bark of trees. Their feet are well adapted for this, having two backward pointing toes to provide the necessary support. Woodpeckers are not easily seen in the wild, for they are rapid flyers and dart quickly from tree to tree, starting near the bottom and working quickly upwards and around the trunk. They are most easily located by listening for the unmistakable drumming sound they make as they drill into trees for grubs and larvae. They also nest in the trunks of dead or diseased hollow trees, drilling a hole for an entrance and clearing a large cavity inside in which to lay the eggs. Some species are now rare, but one of the most common, the lesser spotted woodpecker of Europe is shown here.

19

◀ **Magpie-lark** *Grallina sp.*
The magpie-lark is neither a magpie, which is related to the crows, or a lark, which is a bird of the northern hemisphere. A better name is the mudlark, which comes from this bird's habit of building a relatively large, strong nest of mud reinforced with grass, feathers and other materials. There are four species of mudlarks, and their distribution is limited to New Guinea and Australia. This picture shows two young magpie-larks about to feed. They are wide-ranging feeders, their diet consisting of insects such as locusts, grasshoppers and flies, earthworms, grubs, snails, cutworms and ticks which they remove from the skin of domestic animals. Adult magpie-larks breed near lakes and rivers, and continue to breed as long as it remains wet enough for there to be mud available for nest-building. Two to five eggs are laid, and they are incubated by both parents. In dry years, this bird may not breed at all. Adult males and females form pair-bonds which last for life. The young are also fed by both parents and when they leave the nest and begin to forage for themselves, they join large flocks of other immature birds. The adults immediately begin to raise another brood.

▶ **Secretary Bird** *Saggitarius serpentarius*
This large bird is a native of the southern half of Africa where it lives in open grassland and areas which are not too densely covered with vegetation. Surprisingly, perhaps, they are thought to be most closely related to the falcon family, although their beaks are not as well-developed, nor do they have powerful crushing talons. They are best-known for their habit of killing and eating snakes, sometimes quite large ones. Although these do not necessarily form the major part of their diet, some individual birds seem to prefer reptiles such as tortoises and snakes. Although they are not immune to snake venom, fatalities are rare, and many farmers try to encourage secretary birds to live nearby to keep the snake population down. They also eat small mammals, large insects such as locusts, lizards and young birds. They were first imported into Europe in the eighteenth century, and were given their name because of the black head feathers which reminded people of clerks who carried their quills stuck in their wigs.

▼ Flycatcher *Sigelus silens*
There are more than 100 species of flycatcher, found throughout the world except in the Americas. In appearance, plumage and some of their behaviour they are a very diverse group, but they do have certain characteristics in common. They are woodland birds, usually solitary or in pairs, nesting in crevices, trees or woodpecker holes. As their name implies, they tend to feed mostly on winged insects, which they catch in flight. This kind of precision work is aided by a growth of bristles around the mouth which acts as a sort of net to help capture the prey. Some flycatchers are polygamous, the male accepting little responsibility for the chicks, although in the fiscal flycatcher pictured here, both parents undertake the feeding of the brood. If one parent dies, the other brings twice as much food to compensate.

▶ Robin *Erithacus rubecula*
The robin was originally named the redbreast, but in the fourteenth or fifteenth century the name robin was added. The shortened version, dropping the redbreast, is the one with which most people are familiar. It is interesting to note that the robin must have been a well-known and well-loved bird, for explorers and settlers around the world have applied the name robin to local redbreasted birds although they are not the same species as the British robin. The robin has several noteworthy features. It is impossible to tell the difference between male and female, except for behavioural differences during courtship and mating. Both male and female hold and vigorously defend territories and sing brilliantly throughout the year, except for a short period in June or July. The robin is also noted for nesting in strange places such as pigeonholes and brickwork.

▼ **Ptarmigan** *Cagopus leucurus*

The ptarmigan ranges over most of the northern hemisphere, in rocky regions with little vegetation and a cold, severe climate. It is found in a few other areas in the northern temperate zone, but only at heights well above sea level, regions which more or less duplicate its more northerly habitat. The white-tailed ptarmigan shown here, the male on the left, is common in North America, Europe and Asia. The ptarmigan moults three times during the year, rather than the more usual twice. From a rather mottled brownish grey in the spring, it changes to a grey with black markings in the autumn. Then, with the onset of winter, the third plumage appears, the familiar brilliant white, except for a small black eye patch in the male, giving effective camouflage against the wintry landscape. Like their relatives the grouse, ptarmigan fly rapidly and close to the ground, staying as close as they can to the landscape which their colouration matches so well.

▲ **Jungle Fowl** *Gallus gallus*
*There are four species of jungle fowl, all of them closely
related, found throughout the warmer parts of Asia
from India and China to Indonesia. The most notable
aspect of the jungle fowl, a member of the pheasant
family, is that it is the animal from which the domestic
chicken was derived, and has the same scientific name,
and many of the same feeding and breeding habits. It is,
in fact, open to question whether or not the chicken was
domesticated once, from the jungle fowl, or separately
from more than one of the four closely-related species.
Jungle fowl, which are ground nesters, are extremely
vulnerable to predators in their natural habitat. They
are very wary birds, seen only rarely, and immediately
disappear into the dense undergrowth if disturbed. They
roost in trees at night, choosing those which offer the
greatest overhead protection from night hunters such as
owls. This almost secretive behaviour probably accounts
for the survival of the species in the wild.*

▶ **Roller** *Coracius caudata*
The rollers get their rather strange common name from
their acrobatic displays during courtship. They indulge
in a spectacular rolling and tumbling sort of flight,
screaming loudly and showing off their brilliant colours
to their best advantage. The roller shown here, the
lilac-breasted roller, is one of the most spectacular of the
many species of this group, which is closely related to the
bee-eater and kingfisher families. Most species of rollers
live in Africa, although one, the European roller, breeds
as far north as Sweden, but its range is now becoming
smaller.

▼ **Rifleman**
The rifleman is a fairly common bird of New Zealand.
It is the best known of the small family of so-called New
Zealand wrens, three species which look like, but are
unrelated to, the true wrens. A fourth species, the
Stephen Island wren, became extinct before the beginning
of the twentieth century, and is an excellent example of
the effects of introduced predators. It is believed that
this ground nesting and probably flightless species was
wiped out by a single cat, the pet of the lighthouse keeper
on Stephen Island, the birds' only habitat.

▶▶ **White throated sparrow** *Zonotrichia albicollis*
This bird is one of the many North American finches
which are called sparrows. This group includes the song
sparrow, the grasshopper sparrow and the lark sparrow.
They are found throughout Canada and Alaska and their
range extends into the United States, although they
are more common in the damper climate of the east than
in the west. They are highly territorial birds with a
notable warning song, although they combine into flocks
during the coldest part of the winter. These flocks
usually migrate during the winter to the southern part
of the United States and Central America.

◄ Peafowl *Pavo cristatus*

*These familiar members of the pheasant family are
originally from India, Burma, Java and central Africa,
but their beautiful plumage and easy domestication have
caused them to be distributed all over the world, by
people looking for living ornamentation for their
gardens. In some places at certain times they have also
been used as a symbol of luxury when roasted and eaten.
The two sexes are known as peahen and peacock,
although the latter name is the common one applied to
both sexes. In fact, it is only the male who has the
magnificent display mounted by this blue or India
peafowl. The female by comparison is drab and almost
colourless, although she can raise her rather small and
insignificant train in display as well. This display is part
of the peafowl's courtship behaviour, although the bird
is known to display to other animals, including people, as
well. Such an act of supposed vanity is probably what
has given rise to the expression 'proud as a peacock'.
They seem to be able to eat almost anything, and breed
readily, nesting on the ground or in trees when there is
danger from predators.*

►► Waxbill

*This small, colourful songbird is well-known as a cage
bird. Most of the several species of waxbills live in
Africa south of the Sahara desert, except for one Asian
and one Australian species. Waxbills have been
introduced to Portugal and to South America, where
they appear to thrive in the wild. They are seed and
grain eaters, and although they form into flocks outside
the breeding season, there are not enough of them to
become an economic pest. The female waxbill builds
the nest, with a little assistance from the male. The
finished nest incorporates a so-called 'cock nest',
thought to be either a perch for the non-incubating
waxbill or a decoy for would-be predators. Both parents
feed and rear the young.*

►►► Tanager *Tanagra cyanocephela*

*The tanagers, of which there are more than 200 species,
are entirely confined to the Americas. All but five species
live in tropical zones. These brilliantly coloured birds
live in forests, and are rarely seen. Only a few specimens
of some species have been seen or collected. In most
species, the male and female exhibit the same colouring.
The migratory inhabitants of North America are the
exceptions to this general rule. Tanagers eat fruit,
seeds and insects, and have been known to raid wasps'
nests for larvae and pupae. Eggs are incubated by the
female, but the male assists during the incubation by
feeding his mate. The young are fed by both parents, who
occasionally feed young in other nests as well.*

◄ **Gyrfalcon** *Falco rusticolus*
Very similar to the peregrine falcon, the gyrfalcon is the largest, and some say the most beautiful, of the falcons. It is confined to the areas in and around the Arctic circle except for a few isolated populations found in high, cold mountainous regions in Central Asia. They feed on any available Arctic birds and mammals such as ptarmigan, lemmings, rabbits, mink, weasels and geese. Near coastal areas, common inshore birds such as gulls and ducks make up a considerable part of the gyrfalcon's diet. The nest is usually built on a rocky ledge or in a crevice on a cliff face. It is a crude affair of branches and twigs lined with mosses and grass. Gyrfalcons will use the abandoned nest of another species if one is available. The pair bond is strong, the male and female remaining together for life and continuing to use the same nest year after year. Four eggs are laid, and the female does all the incubating. The male does most of the hunting, turning over his catch to the brooding female who tears it up and distributes it to the chicks. The young are fledged quickly, to take advantage of the later-hatching young ptarmigan.

▶ **Barn Owl** *Tyto alba*
There are ten species of barn owl distributed widely throughout the world. Their numbers became very depleted due to quite unjustified persecution by gamekeepers. After making a small recovery at the beginning of this century, their numbers are once again rapidly declining due to the increased use of agricultural pesticides which remain as poisons in the small animals which the owls eat.

They are sometimes seen during the day but usually do not come out to hunt until twilight. Each individual bird has a regular route that it follows each night when searching for prey. Although it has extremely acute sight, the barn owl does not depend exclusively upon its eyes for hunting. The ears are also highly developed and small flaps behind them enable the bird to detect very accurately the direction of a sound and so judge the exact position of its prey.

Between four and seven eggs are laid and they are incubated by the female alone. The chicks leave the nest about ten weeks after hatching to find their own territories where they remain for the rest of their lives.

▼ **Vireo** *Vireo olivaceous*
There are about 40 species of vireo, found only in North and South America. In North America they are migratory, coming from the south to the eastern United States and Canada to breed in the summer. They are not brightly coloured birds, even in the tropics. The nests are built in bushes or low trees, and are characteristically deep cups of grass and other vegetation bound together with spiders' webs and similar materials. The one in the picture, occupied by a red-eyed vireo, one of the North American species, is typical. Most of the building is done by the female, who also does most of the incubating. The feeding is done by both parents, and the diet consists of insects, insect larvae, small fruits and occasional spiders. Vireos hunt in the trees and shrubbery, and do not appear to hunt much on the ground, except for the red-eyed vireo, which has a taste for snails and slugs.

▶ **Stork** *Ciconia ciconia*
A large white stork its beak filled with nesting materials, its feet and legs braced, comes in for a landing on top of the tree on which it is building its large, solid nest.
Its long, fragile-appearing legs are typical of birds which hunt for food in shallow water. Stork populations appear to be declining across the world. Probably some of this is due to predation, for many are shot while migrating. The town-nesters have been discouraged by polluted air and water. Large areas of land have been drained and cleared also reducing their habitat, and they seem less well-suited to the increasingly wetter, cooler summers of Europe. They may also have been affected by pesticides, although this is not known for certain, but there does not, at present, seem to be any cause for alarm at this decline. Some towns and cities in Europe are now trying to encourage storks to return.

Starling *Sturnus vulgaris*
Although there are more than 100 species of starling,
which are found in Europe, Africa and Asia, by far the
best-known and most wide-spread is the common
starling pictured here. Their numbers have expanded
considerably during the twentieth century, and they are
now commonly found in most kinds of temperate and

sub-tropical environment. They are most noticeable in
European cities in the evenings, when flocks of thousands
wheel round the sky and settle in trees and on the ledges
of buildings. This flying and settling, then flying again,
may last for an hour or more, accompanied by a noisy
chorus of not particularly attractive calls, until the birds
settle in the city for the night. They have favourite

roosting areas in cities, and return to them every night throughout the year except during the short breeding season. On the whole, they are a noisy, messy nuisance.

They have been spectacularly successful in colonizing North America. Several attempts to introduce them failed, and then in 1890-1891, 100 birds were released in Central Park in New York. In less than 60 years, they had spread across the continent to the Pacific, south to Mexico and north through Canada. They are now established in countless millions, and provide more nuisance value than anything else. In both Europe and America they have unfortunately occupied breeding grounds of other species, further enhancing their expansion.

▶ **Marabou** *Ceptoptilus crumeniferus*
*The marabou is a common bird of Africa in the region
south of the Sudan and north of Rhodesia. This
photograph is of an African marabou, accompanied by a
pelican, and there are also closely related marabous in
southern Asia and Borneo. The marabou is one of the
storks, but it also has some of the characteristics of
vultures, with which it shares its range. The almost
featherless head and neck are adaptations for carrion
feeding. It is not uncommon, in East Africa, to find
marabous, vultures and hyenas all feeding on the same
carcass. Like the vultures, marabous soar high on the air
currents, keeping a watchful eye for dead or dying
animals. Unlike the vultures, however, they also
frequent farms in search of dead domestic animals. The
marabou's heavy, powerful beak is capable of coping
with the problems of tearing apart large dead animals,
but it is also used for the rather more stork-like pursuits
of fishing in shallow waters. The long legs are typical of
wading birds. Marabous breed in the dry season, when
their aquatic prey are concentrated in the shrunken,
shallow water holes, and when there is a higher rate of
mortality amongst land animals. The nest is built of
twigs and sticks, and when nest building has begun, one
of the pair always stands guard to protect it from
neighbours who would otherwise dismantle it, using the
materials for their own nests.*

▶ **Harrier** *Circus sp.*
*The harriers are widely distributed across the globe, in
open country. They are found in all continents except the
polar regions and are not present on some islands in the
Pacific. The typical long legs, large head and wide
wing-span of the harriers show particularly well in this
photograph of a harrier coming in for the kill. They are
low-flying birds, a relatively uncommon feature in the
hawks and falcons. They are also atypical in that they
sometimes nest in colonies, and outside the breeding
season form into large flocks for migration, when they
leave the temperate zones in which they breed and move
to sub-tropical regions. Harriers hunt systematically,
making low-flying runs across their territories in regular
patterns which they seem to repeat. They will occasionally
take flying birds, but are more apt to take eggs or chicks
from the nest of ground-dwellers. Small mammals of the
locality are also a favourite prey. The harrier's nest is
built of various grasses on the ground, on a low hummock.
The location is usually chosen with care, for these birds
frequent wet or marshy areas, and the nests could be
susceptible to flooding. Three to five eggs are laid, the
female being responsible for the incubation while the
male continues to hunt for both of them. After hatching,
the chicks remain in the nest for a week or more before
starting to explore, but continue to return to the nest at
night for about another two months, when they become
independent.*

◄ Hummingbird

The hummingbirds are confined solely to North and South America. There are more than 300 species of hummingbird, and some of them are the smallest living birds. The so-called giant hummingbird is only about 20cm long, while the body of the bee hummingbird of Cuba is only the size of a bumble bee. Many of the species are brightly coloured, as is typical of many tropical birds, but in fact the hummingbirds are not confined to the tropics. Some species migrate north to breed, reaching Alaska in the west and New Brunswick in the east. The eastern species migrate across the Gulf of Mexico. The secret is yet to be revealed of how these tiny birds can make this crossing when it is known that they cannot store enough food reserves to provide the energy for a flight of this length. There are also many non-migratory species in Central and South America. Some species are found as far south as Tierra del Fuego. Hummingbirds get their name from the sound made by the beating of their wings, which is so rapid that it produces a hum. The wing beat may be as high as 80 per second. This speed, combined with a remarkable anatomical modification which allows the hummingbird to move its wings through an arc of 360° at the shoulder, allows the bird to dart to and fro rapidly and to hover like a helicopter. The hovering hummingbird, shown in this photograph, is drinking nectar from the base of a flower. Its long curved bill is well adapted for this purpose, and the sweet nectar forms the major part of the birds' diet, although they also catch spiders and flying insects.

► ▲ Trogon *Pharomachrus pavoninus*

The trogons are little-known birds of the tropical forest. There are about 35 species in all, most of which are found in the Americas, although others live in Africa, India, China, the East Indies and the Philippines. Like woodpeckers, parrots and some other birds, they have an unusual toe arrangement, with two facing forwards and two backwards, well shown in this photograph of a South American pavonine quetzal. Trogons feed on insects, which are taken from the foliage while the birds hover in flight, and on other small animals, even frogs and lizards, found amongst the vegetation. South American trogons also eat a lot of fruit.

► ▼ Shrike *Laniarius atrococcineus*

There are several species of shrike, and they are found throughout the tropical and sub-tropical regions of the world, migrating into temperate and sub-temperate parts of the northern hemisphere for breeding. This is a photograph of the crimson-breasted shrike. Shrikes are curious birds, for although they are passerines, or song-birds, they have many of the characteristics of birds of prey. Their diet consists of birds, small mammals, insects, lizards and even the occasional bat. They have the unique habit of impaling their prey on a thorn or similar structure, which assists them in tearing the prey apart with their feet and bills. In some cases, prey may be hung up for storage, as it were, the shrike returning to it days or even weeks later.

◀ **Drongo** *Dicrurus adsimilis*
*There are about 20 species of drongos, natives of the
tropics from Africa through southern Asia to
south-eastern Australia. With their glossy black
iridescent plumage and their long tails they are easily
identified as they hunt or sit in exposed places sunning
themselves. They are one of the few groups of birds whose
ancestry and relationships are quite unknown. It has been
speculated that they are related to the crows, because
they are black, to the shrikes because of the shape of their
bills, and to the flycatchers because of their feeding
habits. No definitive answer has yet been produced. The
common drongo of Africa pictured here, is a fairly
typical member of the family. Their principal food is
insects, and a marked preference for bees and wasps has
been noted. Presumably some mechanism exists which
keeps them from being stung. A small, fragile nest is
built high in a tree in the dense forest, and the two to
five young are reared by both parents.*

▼ **Robin with Chicks**
*The robin frequently nests in strange places, often in
close association with man. This robin has built its nest in
a garden shed and seems quite unruffled by the seemingly
incongruous setting. Five or six eggs are laid in the nest,
and the female is responsible for all the incubation. The
male feeds her during this period, which usually lasts
about two weeks. The male and the female both help to
rear the young, and they may raise two or even three
broods in a season. The robin is an insect eater, but will
also take other common invertebrates such as worms.*

▶ **Turaco** *Tauraco livingstonii*
*The turacos are a diverse group of about 18 species which
live in Africa south of the Sahara desert. The photograph
shows a typical iridescent turaco known as
Livingstone's turaco. Most are about the size of a pigeon,
and all but one species show the typical crest. They seem
to spend most of their time in trees. When disturbed or
frightened, they freeze and when threatened they run
further up the tree into protective foliage. A curious
adaptation which allows one toe to face either forwards or
backwards as necessary gives them additional mobility
when fleeing or gathering food. The turacos are fruit
eaters, but they are messy eaters and drop more than they
actually eat, although the dropped fruit probably serves
to feed some ground-dwelling animals. They will
occasionally eat the tender parts of young plants, and
some species will also feed on any available small
invertebrate. Only two eggs are laid in a flimsy nest, and
although two breeding seasons a year have been
described for some species, it seems likely that each
individual bird breeds only once in a year.*

45

▼ **Oxpecker** *Buphagus africanus*
Although its plumage is drab, the oxpecker's bill is usually brilliantly coloured, as it is in this appropriately named yellow-billed oxpecker. There are two species of oxpecker, both natives of the African plains. They take their common name from their habit of feeding on ticks, flies and other insects which they find on the skins of the large African ungulates. As they are also known to aggravate open sores and wounds and to drink the blood of their hosts, they are not always beneficial.

▶ **Martial Eagle** *Polemaetus bellicosus*
The martial eagle is the largest of all the African eagles. It is found in all regions south of the Sahara desert except for thickly forrested areas. It is a solitary bird, and avoids man and any signs of civilization. This is probably to its advantage, as some people accuse it of preying on livestock. However, this is not proved, and its diet seems to consist primarily of common birds and mammals, young impala and jackals, as well as occasional snakes and lizards.

▶ ▶ **Turtle Dove** *Streptopelia turtur*
The turtle dove is well-known in Europe, for its breeding range extends from North Africa to northern Scandinavia and from Portugal well into Russia. In the winter it migrates to the tropical and sub-tropical regions in Africa and Asia. The turtle dove is primarily a seed-eater, and the bulk of its diet consists of weed seeds, so contrary to popular opinion it is not an economic pest in most regions. It will also eat small ground-dwelling invertebrates. One of the most interesting aspects of the ecology of the turtle dove is its ability to survive even against seemingly overwhelming odds. Although two or three clutches may be raised in a season, any shortage of food will mean that the young will not be able to build up a sufficient food store to survive the long migration. Turtle dove nests are also particularly vulnerable to predation. Magpies and jays are the most serious nest robbers, and both young and adult birds are common prey for sparrowhawks. Young birds are also less successful on migration because of their inexperience. They tend to fly regardless of weather conditions, resulting in large numbers being blown out into the Atlantic Ocean.

▶ ▶ ▶ **Crowned Eagle** *Stephanoaetus coronatus*
The appropriately-named crowned eagle is native to Africa. Although smaller than the martial eagle, it is a more powerful bird. It is found in all parts of Africa south of the Sahara desert except for the extremely dry regions of the eastern and southern parts of the continent. Unlike most other eagles, it likes the dense foliage of forrested areas, roosting in there and hunting just over or under the tops of the tall trees. Prey consists almost entirely of mammals such as small forest-dwelling antelopes, monkeys and others of almost the same size. Birds and reptiles make up only a small proportion of the diet. Nest building is a task taken on by both members of the pair. The male collects the materials and delivers them to the female, who builds the nest. The same pair will use one nest year after year, and when that nest is abandoned, it will be taken over by another pair of crowned eagles. As some nest building goes on each year, in time the nest becomes an enormous structure, near the top of a tall tree. Only one young is usually reared, and it does not become fully independent for at least 18 months.

Bats

Bats are the only truly flying animals which are not birds. Of the more than 4000 recognized species of mammals, more than 1000 are species of bats, making them much more common than is usually supposed.

The wings of a bat are composed of a tough, leathery membrane derived from the skin, supported by the elongated second to fifth fingers of the hand. The hind part of the membrane is attached to the hind legs, and in some bats the tail is included in the membrane. The first digit of the hand is free of the membrane, and has a small claw allowing it to function as a grasping organ. The hind legs are small and of little use, except as grasping tools when the bat is hanging in its roost. Locomotion on the ground is thus obviously severely restricted, and virtually useless. Compared with most birds, bats are not particularly good flyers. They have to work very hard to keep aloft.

The bats' radar system is well known. It is a biological sonar system which allows the bats to fly at dusk or in the dark without crashing into things which they cannot see. This system of using echo-location to pinpoint obstacles is in principle simple and straightforward, although a complex system of connections in the brain is necessary to make sense of the information. As the bat flies, it emits a series of ultrasonic clicks. Like all sound waves, these echo off the surrounding objects, and the reverberations are picked up by the bat's highly sensitive ears. The direction and strength of the signal tells the bat exactly where the object is located in relation to its own position. Thus even a blind bat can fly without any problems, but if it is unable to emit its high-pitched noises, or if it is deaf, it will blunder into anything in its path.

In most bats, except for those which are active in the daylight and do not use echo-location, the eyes are correspondingly reduced, small and not very light-sensitive. Thus the expression 'blind as a bat', while anatomically and physiologically true, is behaviourally inaccurate, because the bat has a highly-developed alternative which gives it extremely sensitive 'sight'. Echo-location gives a very accurate navigational system, and bats have been observed to pass without injury between the blades of fans rotating at several hundred revolutions per minute.

Pipistrelle *Pipistrellus pipistrellus*
This amazing photograph, of a bat in full flight, is of the European pipistrelle, or common bat, the smallest and most common of the European bats. It is not a particularly good flyer, having a rather jerky movement, which probably gave it is ancient name of flying mouse. There are over 40 species of pipistrelle bats, found across Europe, Asia, the Far East, Australia and North America. Its habitat is varied, in appropriate nooks and crannies on buildings, as well as in hollow trees and crevices in rocks. It may live singly or in colonies of several hundred. It feeds mainly on flying insects—gnats, beetles and the like—which it catches on the wing on its nightly flights which begin before sunset but may even take place in full daylight.

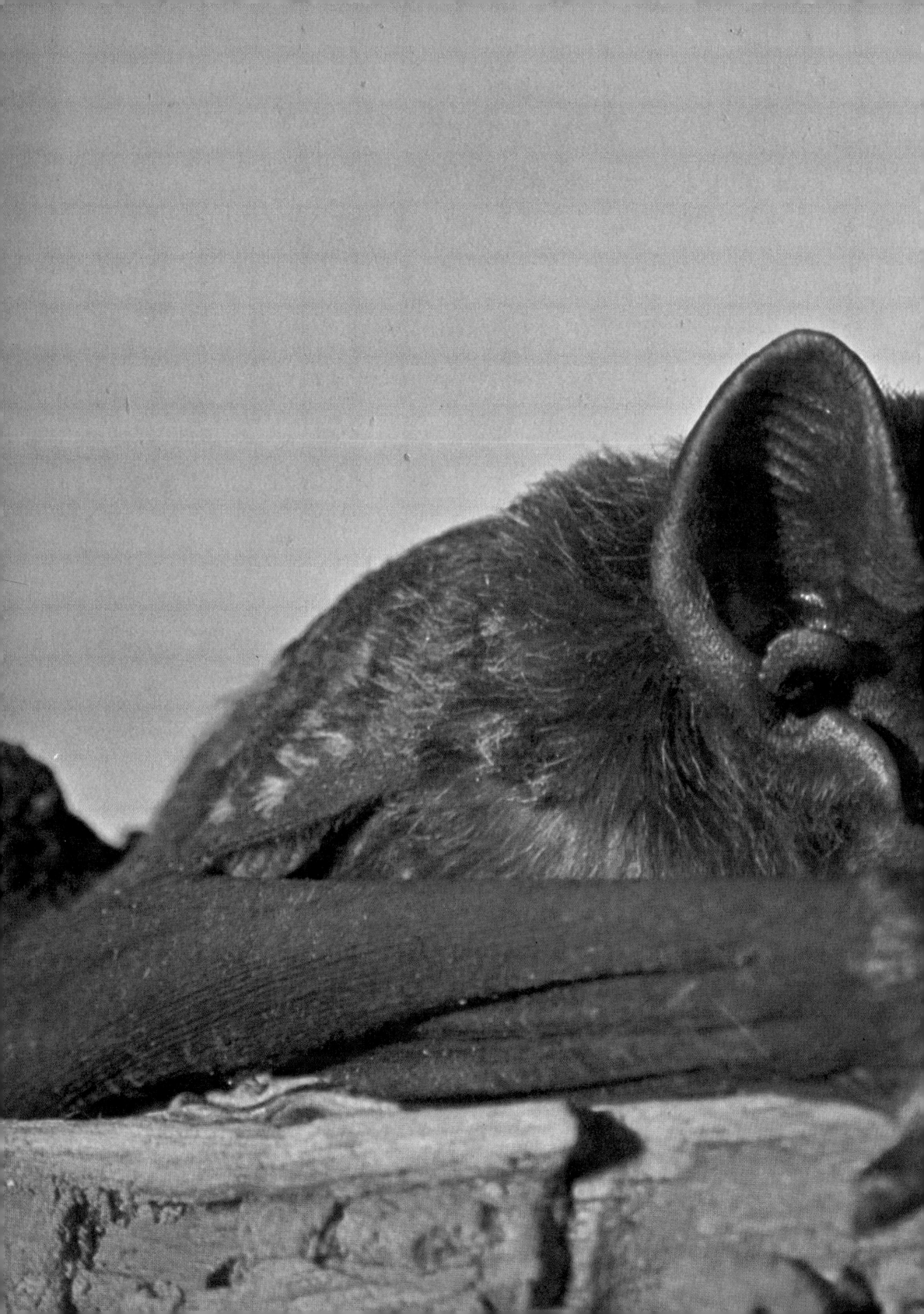

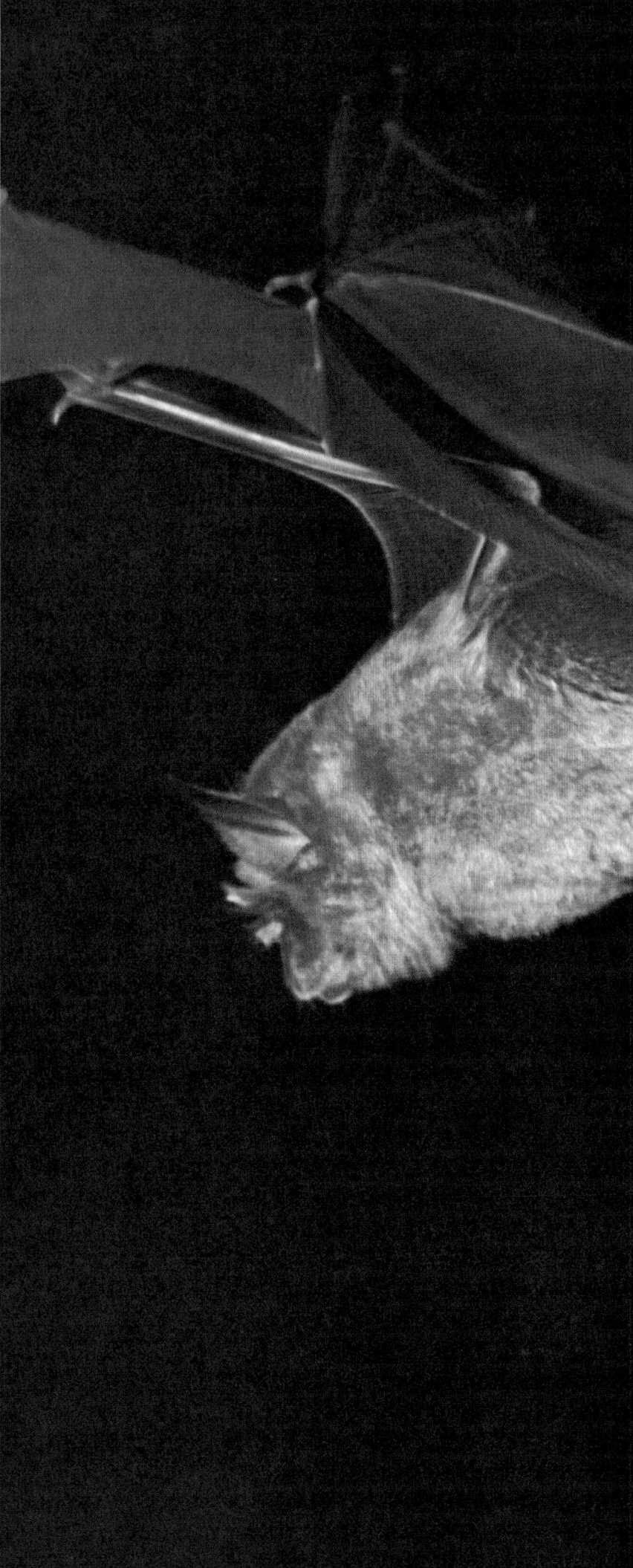

◄ ◄ Tomb bat *Taphozous saccolaimus*
*There are about 40 species of tomb bats which are natives
of tropical and sub-tropical regions of the southern
hemisphere, in Central and South America, Africa, Asia,
the Far East and Australia. Their name derives from
their habit of roosting in the tombs and pyramids of
Egypt, and representations of bats have been found in
tomb wall paintings, suggesting that they have been
associated with these structures for at least 5000 years.
However, it must not be assumed that tomb bats are
exclusively resident in sepulchres for, like all other
animals, they have taken advantage of any possible
dwelling place that they can exploit. Although there is no
need for these warm-climate dwellers to hibernate they
still build up a seasonal store of fat and have a long rest
period during the year suggesting that they may have
evolved from ancestors which did hibernate regularly.
Shown here is one of the species of this group, the small
tomb bat.*

Horseshoe Bat *Rhinocophus ferrumequinum*
*The horseshoe bat gets its name from the horseshoe-
shaped nose-leaves, expansions of the skin surrounding
the upper lip and nostrils. These give the bat its
characteristic appearance and give the impression that
the specimen on the far right, a greater horseshoe bat, is
smiling for the camera. The elaborate structure is thought
to be associated with the bat's echo-location apparatus,
being used to direct the sounds which this species emits
through the nose. There are more than 50 species of
horseshoe bats, which are found in temperate and tropical
regions. Some horseshoe bats are quite large, with wide
wing spans. They fly late at night, making several
journeys each night. They are low flyers, skimming the
ground as they glide along propelled by occasional heavy
strokes of their wings. By day, they rest, either singly or
in colonies, hanging freely, head down, with their wings
wrapped around them giving them a bundled-up
appearance. Males and females roost in separate places,
favourite roosting spots being caves or tunnels and dark
spaces in buildings such as lofts or roof spaces. Although
horseshoe bats hibernate from October to March, at
least in the less tropical parts of their range, they are not
continuously asleep for that period. They are very
sensitive to changes in temperature. If it drops, they
huddle together for warmth. If it rises, they may go out
for a short flight feeding mainly on ground-dwelling
insects if the time of year is not right for there to be
sufficient flying prey. After these little forays, they
return to hibernation, but not necessarily to the same
roost. Females take very young offspring with them on
hunting trips, the single young bat hanging on to a false
abdominal nipple, but later, when the baby becomes too
big, she hangs it up on its own in the roost while she goes
out to forage.*

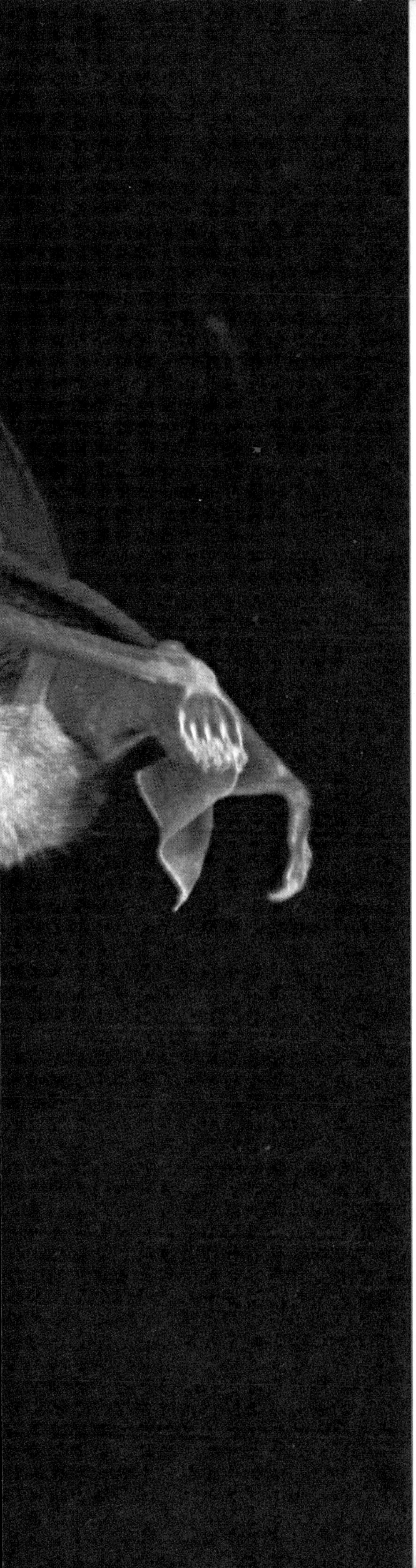

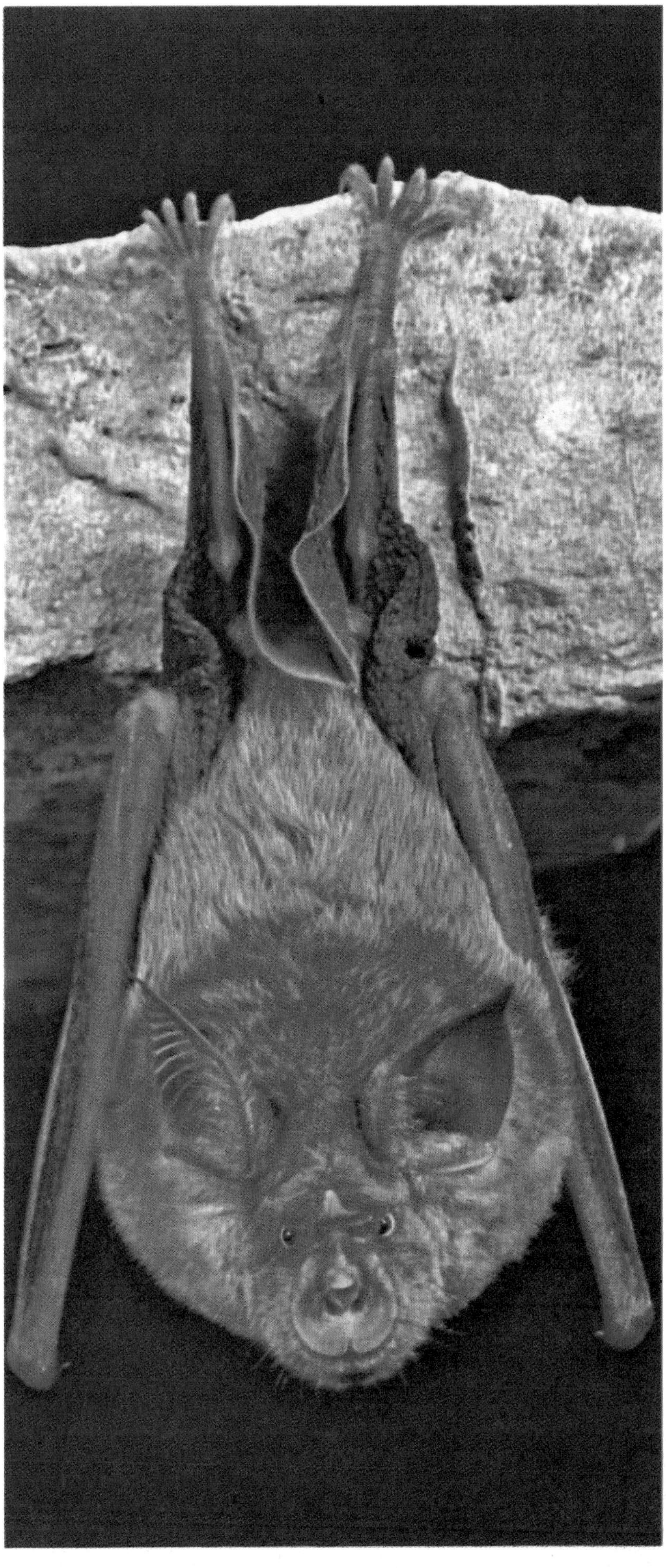

▼ Flying Fox

The flying fox is not a fox at all, but a bat with a fox-like head. There are more than 60 species of flying foxes, forming one of the largest groups of fruit-eating bats. The flying foxes are also some of the biggest bats, with long bodies and a very wide wing span. They are natives of southern Asia, India, the islands of the Indian Ocean and as far east as the Philippines and the north, east and west coasts and adjoining inland areas of Australia. Because of their diet, they do not use echo-location, but have good eyes and a well-developed sense of smell. They can sometimes cause considerable local damage to fruit crops.

▼ **Vampire Bat** *Desmodus rotundus*
*The vampires are probably the least attractive of all the
bats. Contrary to popular opinion, the three species of
vampires are among the smallest of the bats. They are
found only in Central and South America. Vampires
feed on the blood of other mammals, and the adaptation
which facilitates this rather repulsive characteristic is
easily seen in this close-up of a common vampire. The
over-sized, razor-edged upper incisors are well adapted
to allow the bat to open gently a small wound from which
it sucks the blood through a tube formed by the tongue and
the lower-lip. Vampires rarely attack man, but cause
considerable damage to domestic animals.*

Water birds

Water birds can be grouped roughly into those which frequent fresh water and those commonly associated with the sea. Birds common on fresh water include the familiar ducks, geese and swans, as well as less familiar but still well known birds like flamingos and king-fishers. Marine birds are usually well known, although some people tend to call any shore bird a gull. As well as the familiar gull, there are the terns, pelicans, puffins, penguins, and so on. Obviously this sort of classification is not adequate even in a general way. For example, it is well known that geese usually breed in the north, near salt water, and that gulls are common on large inland bodies of water. Many other species appear to be similarly at home on salt or fresh water, although overall most species show a distinct preference

for either one type or the other.

Water birds show many of the same behavioural traits as land birds. They have breeding territories and most species are migratory. Their territorial behaviour is perhaps not as immediately obvious, as many species are colonial nesters and share common feeding grounds, especially marine birds. Many water birds have the most spectacular migratory flights of any birds. Some species of tern, for example, cover many thousands of kilometres. The Arctic tern has the longest migration of any bird, covering up to 40,000 kilometres every year moving to and from its breeding grounds. Although certain landmarks must be encountered during these flights, long distances are travelled over open sea, and it is still uncertain how

Teal *Anas punctata*
The teals, of which there are several species, are found in most temperate and tropical regions of the world. These blue-billed Hottentot teals are sharing the rocks with the larger red-billed Cape widgeon. The teals are closely related to the well-known mallard ducks, and share many of the same characteristics. Most species of teal are migratory, and they are generally strong flyers, although some species seem to try to avoid flying whenever possible and one or two species are flightless, or almost so. Their feeding habits are also duck-like and only a few are divers. They live on seeds and plant growth during the winter and aquatic invertebrates in the summer. Teals nest on the ground, and both parents help to raise the eight to ten young.

these birds navigate. Many theories have been advanced. The most likely explanation seems to be that the birds use the position of the sun, combined with a sense of time, to navigate. It is also possible that other celestial bodies provide points of reference for night flyers. Probably both geographical and celestial navigational points are used, but this does not fully explain how young birds, on their first migration, unaccompanied by adults, can find their way faultlessly to their wintering grounds.

Water birds have certain characteristics which separate them from other birds, although of course, not all water birds show all these adaptations to life on or near the water. In the fresh water birds, the most common characteristic is the webbing between the toes of the feet, an adaptation for swimming. They also tend to have elongated necks which help them to reach food under water. Some are powerful swimmers, both on the water and under it. This adaptation to a swimming type of locomotion is facilitated by adapta-

tions to wing structure, which in turn reduce the birds' ability to fly. Many strong underwater swimmers have to have a long runway to get airborne. They need to run along the surface of the water for a considerable distance, with vigorous flapping, before getting aloft. Feather structure is altered in many water birds, to reduce drag when in the water, and all of them have a well-developed oil gland at the base of the tail. By squeezing this gland with the beak, oil is forced out through ducts. The bird then covers its feathers with the oil by preening, and this provides an effective, light-weight waterproof coat. Birds in detergent-polluted waters lose this film of oil, and may drown as a result. The sea birds of the Antarctic are some of the most highly adapted aquatic birds. The albatross spends nearly all of its life at sea, soaring on currents of wind for hours or days, dropping to the water only to feed. The penguin is so highly adapted to swimming that it has lost the ability to fly, and is shaped more like a fish or a seal than a bird.

▶ **Stilt** *Himantopus himantopus*
*The stilt has the obvious characteristics of a wader —
long legs, extraordinarily long in the case of the stilt, and
a long pointed beak. Although not a commonly-known
bird, it has a wide range, breeding in Africa, Europe,
Asia, North and South America, Australia and New
Zealand in the temperate and tropical zones. It is known
locally by a variety of common names, but in fact one
species has this wide distribution. The second species is
confined to Australia. The stilt is a bird of both fresh and
salt water, frequenting the edges of lakes and rivers,
marshes, flooded land and any other place where there is
an abundant plant growth on which its prey feeds. The
diet consists of aquatic and terrestrial insects in larval,
pupal and adult stages, other aquatic invertebrates,
worms, snails and tadpoles. They nest in colonies, close
to their feeding areas. The nests are built either on the
ground, in which case they are very simple, or in shallow
water, necessitating a much more elaborate structure.
Both parents incubate and rear the three or four young.*

◀ **Oystercatcher** *Haematopus moquini*
*Oystercatchers are a group of waders with an almost
world-wide distribution. This is a South African black
oystercatcher with the typical red bill and legs of this
group, although many species are white underneath.
Their usual habitat is on the shore line, on rocky or sandy
beaches or on mud flats. They are also occasionally found
inland. The oystercatchers feed in the tidal zone, the area
between the low tide and high tide marks, and sometimes
inland. It is difficult to know how they came to be called
oystercatchers, for they do not feed on oysters, which
are not found in their feeding grounds. They do, however,
feed on a variety of aquatic invertebrates including
worms, crabs and several different mussels. They also eat
insects and insect larvae, some plants, and have been
known to raid the nests of other birds for eggs.*

◀ **Swan** *Cygnus olor*

The swan is a familiar bird and the best-known species is the mute swan, shown here. Originally natives of Asia and southern Europe, they have been introduced to many parts of the world —northern Europe, North America and even Australia, where they have gone wild. Other species of swans are distributed around the world. The swans are close relatives of the geese and ducks, and bear many of the same anatomical and behavioural characteristics. Mute swans mate for life, and retain the same nest from year to year. They are highly territorial birds, and the males aggressively attack any intruders. The nest is built where it is surrounded by water, in reed banks or on small islands, although swans in urban areas are not so particular, and build on land occasionally. There can be as few as five or as many as a dozen or more young.

▼ **Oystercatcher colony**

It is interesting to note the methods which the oystercatcher has developed to cope with the problems associated with feeding on mussels. Limpets, for example, are struck with a sharp blow of the beak, either knocking them off the rock or holing them. With mussels open under water, the bird rapidly inserts its bill in the opening between the shells, attacking first the powerful muscles which the mollusc uses to close its shell. When these are severed, the shells fall apart and the contents are eaten. If the shell is closed, as it is when the mussel is above the waterline at low tide, the oystercatcher carries it to a sandy spot and places it upside down so that the hinge of the mussel, which is the weakest part of the structure, is uppermost. A few hard, quick blows are enough to penetrate the shell, and then the same procedure is followed.

◀◀ Razorbill *Alca torda*
*The razorbill is a fairly familiar bird of the north
Atlantic coast. It spends much of its time at sea, moving
to the coast and further south in the winter with the
onset of breeding. Like most other marine birds, it nests
in colonies, arriving on the breeding grounds as early as
January. A single egg is laid on the ground, in a
relatively sheltered position. After hatching, the young
chick is fed by both parents. At two to three weeks of age,
it makes its way into the sea, frequently by just dropping
off a cliff, and swims out to sea with its parents who feed
it until it is able to fly. The chicks are already
very strong and capable swimmers and they are
thoroughly protected from the sea by their feathers
and a thick layer of fat.*

▼ Pink-footed Goose *Anser fabalis brachyrhynchus*
*The pink-footed goose, which is a variety of the bean
goose, is a native of the north Atlantic, nesting in
Greenland, Iceland and Spitzbergen. It moves south to
Britain, France, Germany and the Netherlands in the
winter, and has been seen occasionally in Russia and
North America. Pairs are formed before the breeding
grounds are reached, and nest building and egg laying
take place soon after arrival. They tend to nest in loose
colonies in relatively inaccessible places, to avoid the
threat of predators, but occasional pairs do nest in open
country. The female incubates the eggs while the gander
stands guard. The young leave the nest soon after
hatching, move to the water, and do not return to the
nest again.*

▼ **Sandpiper** *Actitis hypoleucos*

*There are several species of sandpiper, and the common
name is somewhat confusing as their near relatives,
which include the woodcock, the snipe and the curlew,
closely resemble them, and are frequently mistaken for
them. Like other primarily shore birds, they are
long-legged and sharp-billed. They are most familiar
outside the breeding season, when groups or small flocks
may be seen feeding and running along the shore line,
frequently taking flight, wheeling and landing as a body.
The breeding grounds are in the north, so that their
appearance in temperate areas is during their southward
migration, a sometimes leisurely journey which, in some
species, may be as much as 12,000 kilometres. The
sandpiper shown here is the common sandpiper.*

Black-headed gull *Larus ridibundus*

This gull is a very common bird both on the shoreline and inland. It is one of the smaller gulls and is distinguished by the dark feathers on its head. These, however, disappear during the winter, except for a few small patches. It seems likely that this marking is associated with the bird's mating display.

Gulls are scavengers, although they will hunt for flying insects and small shore invertebrates such as winkles and worms form an important part of their diet. They are also nest robbers and will steal and eat the eggs and chicks of other species and of other members of their own species.

They are colonial birds and several thousands of breeding pairs may be found nesting in one area. The nest is usually an untidy structure and because the colony is very vulnerable to predators the birds do not remain there all the time until it is essential for incubating the eggs. About four eggs are laid but no more than three hatch. Both parents incubate the eggs and take turns in feeding the chicks. They learn to fly when they are about five weeks old and then leave the breeding colony to join the adult birds at the feeding grounds.

Black-headed gulls are found throughout Europe and Asia although they do not range as far north as northern Scandinavia and Russia. Some birds migrate to the Mediterranean in winter and individuals have been found as far south-west as Mexico.

▼ **Pochard** *Aythya fuligula*
Pictured here is one of the several species of pochard, the European tufted duck widespread from Iceland to the Bering Sea. Best known of the pochards, however, is the North American canvasback, a popular sporting duck because it is said to be one of the best tasting. The pochards are birds primarily found in the northern hemisphere, although some are found in eastern and southern Africa, Australia and New Zealand. Like most widely-distributed birds, the pochards are variable in their habits. Some eat only plant materials, while others include considerable numbers of aquatic invertebrates, and even fish and frogs, in their diets. They are mostly found on water, and are awkward and ungainly on land. Many are divers, staying under water for up to a minute, while others are dabblers like their near relatives the mallards. Nests are built near the water in reeds or similar protective vegetation, in marshes, ponds and lakes. From six to a dozen eggs are laid and incubated by the female. The male has little to do with the rearing of the young.

▶ **Kingfisher** *Alcedo atthis*
This well-known but rarely seen bird has an extremely wide distribution, from the northern temperate zones to the entire southern hemisphere except for the Antarctic and the desert regions. There are more than 80 species of these colourful birds, and this photograph of a common kingfisher shows the typical bright plumage. Its behaviour during feeding is remarkable. It perches on a branch over the water until it spots its prey, usually a small fish. When one is spotted, the kingfisher dives into the water, its momentum carrying it through the water with the wings operating like the fins of a fish for propulsion and steering. It seizes its prey and, with a couple of sharp strokes from the wings, the kingfisher surfaces and flies directly back to its perch and has its meal. From this behaviour the bird derives its name, but strangely enough, the majority of kingfisher species, primarily those in the tropics, are not fishermen at all, but feed on land animals and flying insects. They may even turn to nest robbing, and one species, in New Guinea, even digs for earthworms.

◄◄ **Sunbittern** *Eurydyga helias*
The sunbittern is misleadingly named, for, unlike its namesakes, which are members of the heron family, it is more closely related to the cranes. It does, however, look reasonably similar to the true bitterns, and this is doubtless responsible for its name. The reason for the addition of the 'sun' is surely obvious from the picture. The sunbittern lives either singly or in pairs or family groups, in swamps or along the banks of rivers, in Central and South America, from Mexico to central Brazil. Although a good and noiseless flyer, it seems to prefer to spend its time quietly in the shallows, hunting for food, or swimming. They will fly into the trees when disturbed or threatened. Their food consists of insects, crustaceans, small fish and other aquatic animals living in the shallows. Like the herons, the sunbittern stands motionless in the water when hunting, waiting for its prey to come within striking distance and then shooting forward quickly to grab it in its long and pointed bill. Little is known about their nesting behaviour in the wild.

Storks *Ibis levocophala*

Storks are still relatively widespread, although the more popular, wide-ranging species are undergoing obvious decline. The most common European species is the white stork, which figures largely in various legends and has for centuries nested in urban areas, on tops of houses. In various European countries it is a mark of distinction for a family to have storks nesting on the roof, as they are regarded as symbols of fertility, good luck and general prosperity. The white stork is also easily noticed during migration, when flocks of these large birds are commonly seen. The black stork is more common and more widespread than the white, ranging from central Europe across Russia to Korea and Japan. It is also common in the more temperate zones of the Middle East. In the winter, storks migrate to Africa and there are two populations of storks, in Rhodesia and South Africa, which have presumably been derived from migrating birds which chose not to go north again in the summer. Other species of storks are found in Africa, Malagasy, southern Asia and Australia.

The two photographs are of painted storks found in southern Asia. The photograph on the left shows a painted stork coming in for a landing, while on the right, a group of storks in a colony stand on or near their nests at sunset. The nest is a large, solid platform of twigs and sticks, usually at or near the top of a tree. The density of nests in the colony depends on the number of sites on which nests can be built. The nests are re-used year after year. Six to eight eggs are laid by the female, at intervals of two days. They are incubated by both parents for about a month. The eggs therefore hatch at the same intervals, and the youngest birds may not survive if the food supply is in any way restricted, for the older birds will be larger and better able to compete for the food which the parents bring to the nest. Storks usually nest near open water, and their food consists of small fish, frogs, and other aquatic animals which frequent shallow water. They also forage on land, and are efficient and useful predators of locusts, grasshoppers and other large insects. It has been reported that storks have also caught and eaten small mammals, lizards and the chicks of ground-nesting birds. The long, pointed bills are well adapted to the fishing and scooping techniques which these birds have developed for hunting in shallow water.

◄ **Crane** *Grus rubicanda*
There are 14 species of
crane, and except for one
or two, like the European
common crane, they are
now reduced to restricted
breeding areas, and their
numbers are in severe
decline. Two species, the
whooping crane of North
America and the
Manchurian crane of
Japan, are particularly
endangered, there being
only 40 living specimens of
the whooper and 200 of the
Manchurian in a single
swamp in Japan. Cranes
are particularly
susceptible to hunters and
to man-made obstacles
encountered during their
long migrations. The
draining of swamps and
marshes, their traditional
breeding grounds, has
further restricted them.

▶ **Shelduck** *Tadorna
tadorna*
There are only two species
of shelduck, the ruddy
shelduck of north Africa,
Asia Minor, the Middle
East, Russia and China,
and the common shelduck,
pictured here, which lives
in northern Europe, central
Russia and central
China. Both species are
migratory, moving from
their breeding grounds to
semi-tropical and
sub-tropical regions in the
winter. The common
shelduck is a bird usually
associated with salt water,
although in Asia it settles
near large lakes. Its food
is made up almost entirely
of shore-dwelling animals,
primarily snails and other
molluscs, although they
will eat crabs and fish, and
occasionally marine
algae and shore plants.
The nest is built in a hole
or hollow, or even the
deserted burrow of a
mammal, near the shore
line. Eight to twelve eggs
are laid and incubated by
the female while the male
stands guard nearby.

◄◄◄ Terns

The colonial birds like the gannets and the terns are best known from their complex behaviour patterns during mating, egg incubation and the rearing of the young. Like other birds, the colonial nesters exhibit well-developed territorial behaviour, but in this case the territory is greatly reduced. The feeding ground – the sea – is communal ; there is no way in which individual territories could be identified and defended, and in any case, because of the high mobility of their prey, sea birds must be free to roam widely in search of food. The territory usually consists only of the nest and the small area around it. The young birds are safe from attacks from neighbours as long as they remain on the nest, but are immediately attacked and often killed if they venture into an adjacent territory. Colonies are usually mounted by the parents. The adults are able to find their own young apparently without difficulty.

◄◄ Gannet *Sula bassana*

Like many other marine birds, gannets spend most of their lives at sea, coming ashore only to breed. They are found principally in three locations – the north Atlantic, the south Atlantic and in the straits between Australia and New Zealand. Some ornithologists believe that each of these areas is the home of a single species, while others maintain that there is only one species, the common gannet, which exhibits minor geographical variations. A gannet which is hunting and feeding puts on a spectacular show. It flies very high, then dives, wings folded, head first into the water at a great speed. A strengthened skull and a system of air sacs around the brain helps the bird withstand the impact. Only one egg is laid. It is incubated and the young reared by both parents, who abandon it after about two months. The young bird, left to fend for itself, makes its way to the sea, where it learns to fly and fish.

◄▼ Woodcock *Scolopax rusticola*
Woodcock are waders, perhaps surprisingly, members of the sandpiper family but they are unlike their relatives in many ways. They are solitary birds, pairing only during the breeding season and are rarely seen in groups of more than two birds. They are birds of the woodland, although some are found in semi-open country as well. They seem to like damp, marshy areas, feeding on earthworms, insects, and molluscs. Only in bad weather or during times of food-shortage are they found along the shore, the more typical location of waders. This photograph shows a male European woodcock performing his evening 'roding' flight, a patrol of the boundaries of his territory at sunrise and sunset, which establishes the limits and serves to keep out intruders. Other woodcocks present in his territory are attacked and chased away. The European or Eurasian woodcock ranges across north central Europe and Asia from Britain to Japan.

▼ Ibis *Threskiornis aethiopia*
There are almost 30 species of ibis, and they are widely distributed in Europe, Asia and North and South America. They are related to the herons and flamingos, and have many of the same characteristics – long necks and bills and long, spindly legs. Ibises also bear some resemblance to storks and cranes, and many people confuse the various types. The ibises also have another, less fortunate, characteristic in common with their relatives in that several species of ibis are now in danger of extinction. They are now locally extinct in many parts of their former ranges, and the ibis shown here, the so-called sacred ibis is a good example. It derives its name from its historical associations with the priests and pharoahs of ancient Egypt, where it was an integral part of the religious life, the god Thoth having the head of an ibis. The sacred ibis is now extinct in that part of the world, although still abundant south of the Sahara desert.

◄ **Shoebill** *Balaeniceps rex*
The shoebill is a native of Africa, but because of its very secluded way of life, nesting deep in marshes and swamps, little is known about its full range and not much more about its habits. It obviously gets its name from the massive, almost oversized bill, whose usefulness has not been determined although it has been suggested that it might be used for digging large lungfish out of the mud for prey. There is some doubt about the shoebill's lines of ancestry, and some scientists believe it to be more closely related to the pelicans and cormorants than to the storks. One of its heronlike qualities is the ability to stand motionless on one leg.

► **Flamingo**
There are four species of flamingos. The most wide-spread is the greater flamingo shown here. It is found in coastal areas in Central and South America and across central and southern Europe and Asia. There is also an African species, the lesser flamingo, and two species confined to the Andes. Flamingos are colonial birds, flying in large flocks, and the sight of thousands of them taking flight from the shallows of a lake must be one of the most spectacular sights in the animal world. The European greater flamingo is a migratory bird.

◀◀ **Grebe** *Podiceps ruficolis*
*There are 18 different species of grebe, and they are
found mostly in the northern, temperate and drier
sub-tropical regions of the world. Some, like the great
crested grebe, are widely distributed. The great crested
grebe is found in Europe, Asia, Africa, Australia and
New Zealand. Others are more restricted. Ten species
are found only in North and South America, and some of
these are severely restricted, such as the three South
American species which are each found only on one lake.
This is a picture of the little grebe, the smallest of the
grebes, widely distributed through Europe, Asia, the
East Indies, Japan and Africa. Grebes are not ardent
flyers and they have to run along the surface of the water,
flapping as rapidly as possible, to attain enough speed for
take-off. They are much more at home on the water,
where they are good swimmers and powerful divers. They
eat most aquatic animals – insects, crustaceans and
amphibians, along with some vegetation but their
principal food source is fish, which they can pursue
under water with great speed and endurance.*

◄ **Shearwater** *Puffinus gravia*
*There are several species of shearwater, a sea bird which
is a close relation of the albatross. It gets its name from
its habit of flying or gliding for extended periods close
to the surface of the water, feeding on surface-dwelling
fish and other aquatic animals. Shearwaters spend most
of their time at sea, coming on land only to nest and rear
their young. They breed on isolated islands, where they
are free from predators, and along some European and
Australian coastlines. They are colonial birds, nesting in
large groups of up to a few million birds. They are not
well-known birds, for they come to land only at night,
returning to the open sea before dawn. This photograph of
a group of sooty shearwaters was taken just before
sunrise, as they were preparing to take off to sea again.
Only a single egg is laid, and the seven weeks of
incubation are shared by the parents, each of whom may
spend up to two weeks on the nest, living off reserves of
fat built up from intensive feeding just before egg-laying.
The adults desert the young chick before it is fledged,
having fed it intensively during visits to the burrow.*

▼ **Pelican** *Pelecanus onocrotacus*
There are eight species of pelican, including the common white pelican shown here. They are found throughout the tropical and warm temperate zones, on both fresh water and salt water. They are powerful, graceful flyers, using the air currents for soaring, although their size makes the task of getting airborne a difficult one. When they sight food, they dive perpendicularly into the water, sometimes from a great height. Their diet primarily consists of surface dwelling fish but may include crustaceans. Their enormous pouches, which are a modification of the lower part of the bill, are their most obvious distinguishing feature. Contrary to popular belief the pouch is used as a sort of fishing net rather than as a place for storing food.

▼ **Smew**

The smew is one of a group of colourfully plumaged ducks which includes the mersanger and the goosander. It is smaller than its relatives and is found in the extreme north of Europe. The white male has a black line along his back and black patches on his face, while the female is grey with a chestnut head. Their migratory habits depend upon local weather conditions and if the winter is mild, they remain in their breeding grounds. If the weather is very bad they may migrate as far south as the sub-tropics and they are occasional visitors to the British Isles. They are mainly fish-eaters but, because they inhabit estuaries rather than the open sea, frogs and insects supplement their diet.

◄ Swan *Cygnus olor*

Surrounded by water, behind camouflaging reeds, two mute swans repair and renovate their nest. The nest is little more than a large mass of vegetation, which appears to be floating on the water, piled up around a cone-shaped depression in the middle in which the eggs are laid. The nest is usually isolated but a few populations of swans have been found nesting in colonies rather than spaced out individual territories. Swans are bottom feeders, foraging mainly on plants, although they are also known to eat aquatic invertebrates, insects, tadpoles and small fish. Swans are not divers, and therefore they are almost always found in relatively shallow water where their long necks can reach the bottom by dipping or occasional up-ending. Swans are heavy birds, but in spite of this they are graceful if only occasional flyers. Because of their size, they require long stretches of water for take-off and landing, and do not manoeuvre particularly well in flight.

▼ Diver *Gavia immer*

The diver is probably better known as the loon. The four species of loon are circumpolar in distribution, and the most familiar of these, the great northern loon shown here, is a native of Canada, the northern United States, Greenland and Iceland. One of its most distinguishing characteristics is the interrupted ring of white around its neck, called the loon's necklace by the North American Indians. Loon are fish eaters, and they dive and chase their prey under water with incredible speed and agility. They are fast, strong flyers, but require a long run to get off the water. The cry of the loon, a long, gurgling, almost maniacal laugh, is a common sound on northern lakes and rivers. Otters, foxes, mink, crows and gulls prey on the loon and are a particular danger to eggs and chicks. The loon conceals the location of the nest when leaving it, by diving into the water, submerging completely and surfacing further off.

◄◄ Kittiwake *Rissa tridactyla*

The kittiwake is a small gull of the northern hemisphere. There are two species of kittiwake, one a native of the north Atlantic, shown in this unusual photograph, and another, the red-legged kittiwake, which is found in the north Pacific and the Bering Sea. Unlike the other gulls, the kittiwake is a bird of the open sea and nests on cliff faces rather than along the shore line. Except during the breeding season, when they nest in large colonies, they spend their time at sea, in large flocks of 1000 or more birds. They will only come to land in heavy weather, or to breed. A nest of mud and vegetation is built on a rocky ledge. Most gulls do not build elaborate nests, but the kittiwakes must, to keep their eggs from falling into the sea. Kittiwakes feed on aquatic animals but, unlike other gulls, they are not nest robbers. They do not steal and eat the eggs and chicks of other birds.

◄ Greylag *Anser anser*

The greylag goose, one of the five species of the so-called grey geese, is the acknowledged ancestor of the common domestic goose. It is a large bird and like its relatives is a strong flyer, undertaking long migrations from its northern breeding grounds to its southern winter home. Its breeding ground, once extending over most of northern Europe, is now reduced to regions in Iceland, Scotland, Scandinavia and the Balkans. Greylags frequent marshes, and feed on aquatic plants and cultivated land, primarily on grain, rice and stubble. The adults mate for life. Up to eight eggs are laid in a concealed or floating nest. Incubation is by the female, but the gander stands watch to defend her during the incubation period. Both parents help to rear the young, which although fairly independent and capable of feeding themselves from hatching, remain with the parents for up to two months.

▲ Cormorant

Cormorants or shags are found all over the world, except for a few islands in the central Pacific. They are found mostly along the coast line, although some species, most notably the common cormorant, are found on fresh water in many parts of the world. This photograph shows adult and young blue-eyed shags on their Antarctic nesting grounds, along with a few chin-strap penguins on the same island.

Puffin *Fratercula arctica*
There are three species of puffin, all found in the northern hemisphere. Most common is the Atlantic puffin shown here coming in for a landing, its feet braced for steering and breaking, and looking rather awkward and embarrassed. The other two species are natives of the Pacific. Puffins are members of the auk family, but they are sometimes known colloquially as sea parrots, because of their large bills, and are also faintly reminiscent of penguins, with their rather squat shape and waddling walk. Puffins usually feed far out to sea, diving under the water to pursue and catch their prey.

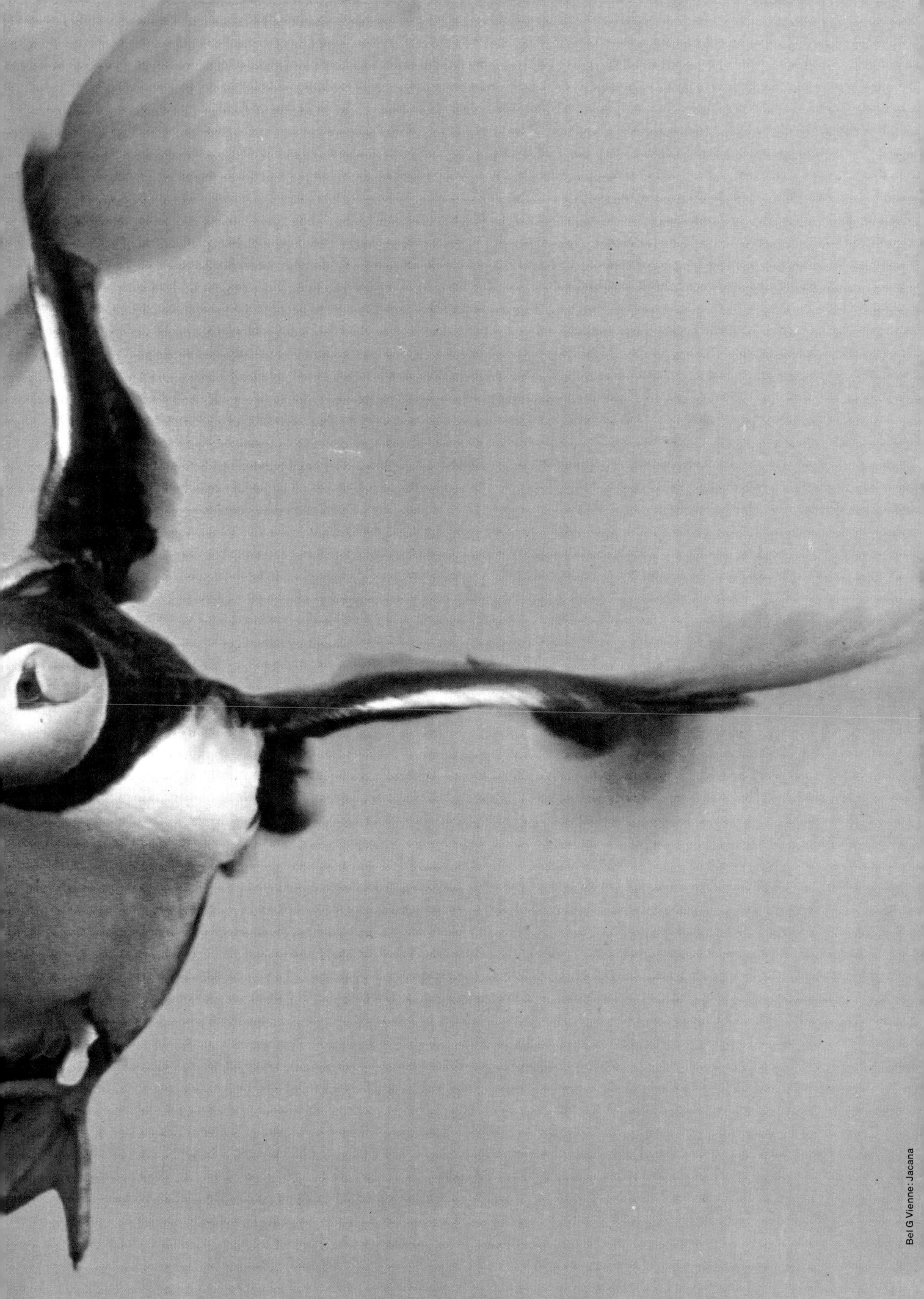

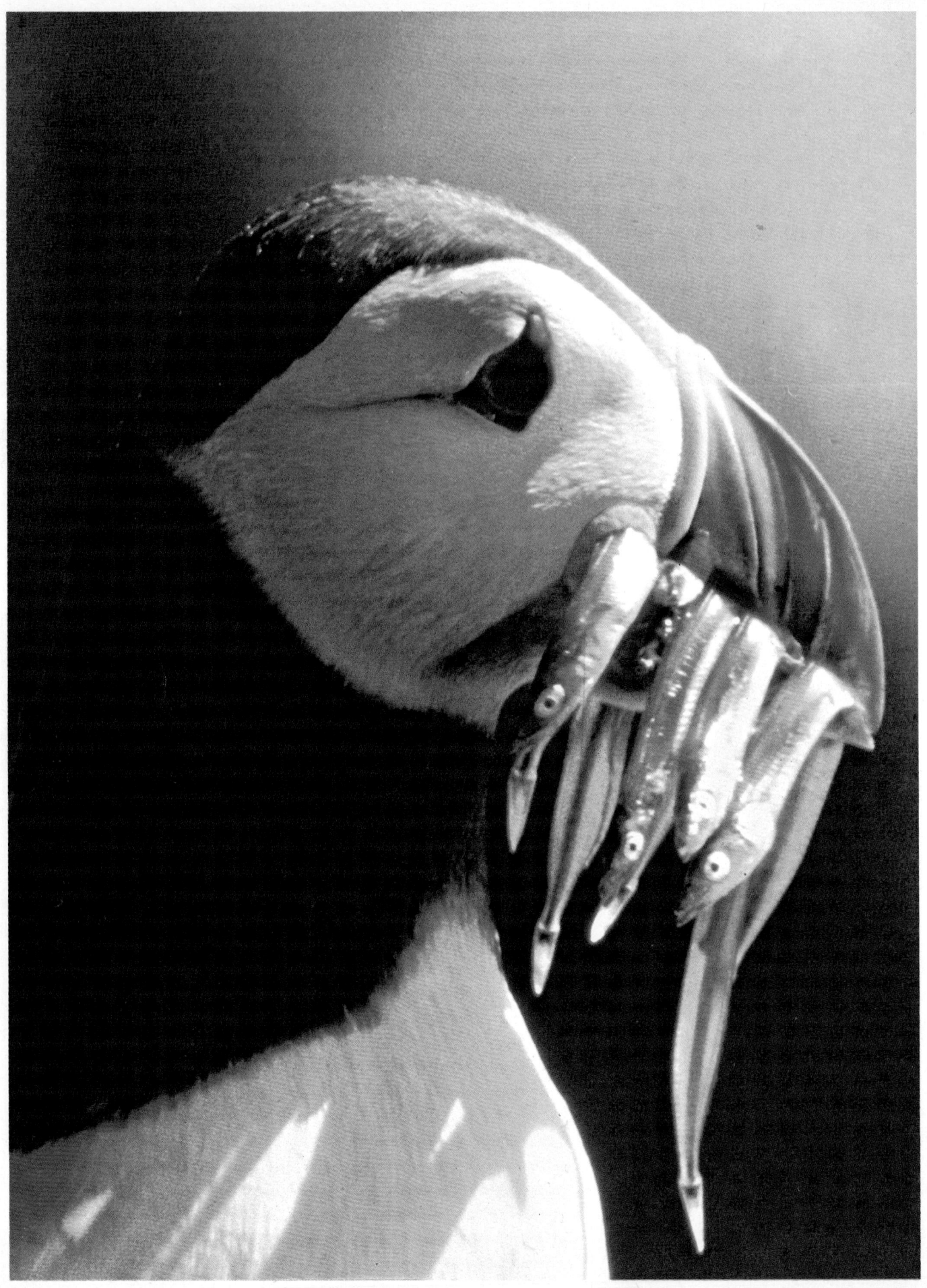

▲ Screamer *Channa torquata*
At first sight, the screamer looks pheasant-like, but in fact is a fairly close relative of the ducks, geese and swans. Shown here is the best-known of the three species of screamers, the crested screamer, a pair of adults accompanied by their chicks. All three species live in South America and inhabit marshes, lagoons and other damp areas. They are not good swimmers.

◄ Puffin
Puffins are efficient fishermen, as this photograph of a puffin with a collection of sand eels shows. When the young are on the nest, the puffin will collect and carry food back. How it can build up a collection like this, adding new fish to the catch without losing its grip on the ones already caught, has not yet been explained. Catches of up to 30 fish have been reported. Puffins are colonial nesters, digging burrows in the earth or taking over empty shearwater or rabbit burrows. Pairing takes place at sea, before nest building, and a single egg is incubated by both parents. The young chick is fed intensively by both parents for about six weeks after which the parents desert the young one and go to sea to moult, during which period they are flightless. The chick remains in the nest for another week, and then leaves the burrow one night, to avoid predators, and makes its way to the sea on foot.

►► Heron *Ardea goliath*
The term heron is a rather unspecific name used to describe a group of birds made up of egrets, bitterns and herons. The many species of herons are widely distributed over the globe, and are common in most places except for South America and Australia. The herons which nest in the northern hemisphere migrate south in the winter. This is a picture of one of the larger herons, the goliath heron, one of the lesser known species and a native of Africa. Most of the herons display the typical characteristics of wader-fishermen – long legs and neck and a long sharp bill. The typical pose of the hunting heron is standing motionless in the water, waiting for fish or other suitable prey to come within striking distance, but some herons have been known to pursue small mammals and birds on the ground, and feed on sedentary aquatic animals as well. In spite of their size, herons are smooth and graceful flyers.

Wildlife of the land

Here we will examine a variety of common and uncommon land mammals and birds. An animal is classified as a mammal if it has both hair and mammary glands. All mammals have hair on their bodies, but no other animals do. The mammary glands are highly specialized skin glands which secrete milk to nourish the young, and are most highly developed in female mammals.

The mammals are believed to have evolved from reptile stock. Their particular advance from the reptiles was the ability to control their temperature,

making them independent, to a very great extent, of the prevailing temperature of the region in which they lived. This high body temperature has enabled the mammals to spread out into all parts of the globe and make use of every possible ecological niche in the water, the land and the air. Birds are also able to control their body temperature, and they have also been able to adapt to a wide variety of environments, although in many ways they are not as highly developed as the mammals, and retain several reptilian characteristics.

Panda

Because the panda's habitat – Tibet and southwestern China – is so inaccessible, not very much is known about its habits in the wild. Attempts to breed pandas in captivity have failed in western zoos and the successes of Chinese zoos were probably produced by artificial insemination. Bamboo shoots form a major part of the panda's diet which also includes other plants and grass. They do eat some small animals such as birds, rodents and fish which they scoop out of streams using their paws like hands.

◄ Woodchuck *Marmota monax*
The woodchuck, or groundhog, is one of the larger rodents and one of the more common mammals of the fields and open woodlands of Canada and the eastern United States. It has a legendary but mythical ability to forecast the end of winter. The woodchuck hibernates all winter but was thought to emerge from its burrow on February 2nd (Groundhog Day).

Mammals have specialized in one other direction, and this has enabled them to become the dominant living group of animals. They have the most sophisticated nervous system of any animal, and they have developed particularly the cerebral portion of the brain, giving mammals great neuro-muscular control and a mechanism by which more varied adjustments and responses to environmental conditions are possible.

There are, broadly speaking, two groups of mammals. Primitive mammals are the monotremes and marsupials. The other, more advanced group is made up of most of the world's living mammals, and it is the advances that these, the so-called placental mammals, have made over their reptilian ancestors which will be considered next.

Hair has already been mentioned as a defining characteristic of mammals. It forms an insulating layer over the body which reduces heat loss. Heat dissipation, necessary in all mammals from time to

▼ Bobcat *Lynx rufus*
The bobcat or American wildcat is one of the medium-sized cats, and is found exclusively in North America. It is a solitary but relatively fearless animal, and will defend itself ferociously if attacked. The female will also vigorously defend her young, even from the male, who has no contact with them until they are weaned. It is closely related to the lynx.

▲ Langur *Presbytis entellus*
There are several species of langur. This photograph is of the entellus or Hanurman langur, a native of India. It is regarded as a sacred animal, and is therefore unmolested and able to live in close association with civilization.

▼ Polar Bear *Thalarctos maritimus*
The polar bear is one of the largest and most carnivorous of the bear family. It lives along the southern edge of the Arctic pack ice throughout the northern hemisphere.

time, is achieved by an increased blood flow through the skin and by the evaporation of water through the sweat glands.

The dentition of mammals is greatly improved and specialized according to the animal's diet. Similarly its digestive tract has been modified to deal with the particular type of food the animal eats. The respiratory system is also improved, with more efficient gas exchange taking place in the lungs, the evolution of the muscular diaphragm to draw air into the lungs, and the separation of air and food passages in the mouth by the secondary palate. The circulatory system is more efficient, blood flow is increased, and the excretion of metabolic wastes, considerably higher in mammals because of the high metabolic rate, is more efficient because of greatly improved kidneys.

Most obvious are the advances in their reproductive methods. All but the primitive mammals give birth to live young which are miniature adults.

Primates

The primates are a large widely-differing group of relatively unspecialized mammals. There are three broad groups of primates. The first of these is the lemurs, of whom the most primitive, the tree shrews, closely resemble the primitive insectivores from which they probably arose. The lemurs and the lorises are believed to represent the next stage in primate evolution, having long tails, flexible limbs and grasping hands and feet. The second group is represented by a single surviving animal, the tarsier.

The third group is the anthropoids—monkeys, apes and man. They are similar in that all of them have large, complicated brains, acute vision and forward directed eyes enclosed in bony sockets. They tend to sit upright so that the hands are free to manipulate limbs for locomotion. Both the thumb and the big toe are opposable in most anthropoids, and this adaptation is well-developed in this group.

There are two groups of anthropoids. The smaller group, the American monkeys, became isolated from the primates relatively early during primate evolution and evolved independently. Their most obvious distinguishing characteristic is their widely-separated nostrils, directed forward and towards the side. This separates them from the non-American monkeys, which have close-together, downward directed nostrils. The American monkeys usually have a long prehensile tail, while their Old World relatives' tails may be long, short or absent, but are never prehensile. The Old World anthropoids can be further subdivided. The monkeys, which include the macaque, the mandrill, the baboon, the langur and others, sit upright, and have bare, hardened pads on their buttocks. The apes, of which the living species include the gibbon, the orang utan, the chimpanzee and the gorilla, have rudimentary tails, arms longer than their legs, opposable thumbs and big toes, and broad flat chests.

How true is the expression 'Man is descended from the monkeys'? The easiest thing to say is that the spirit of the statement is accurate enough, but the actual wording is misleading. It is certainly obvious that man's closest relatives are the other primates, but there is no one animal that can be said to be man's ancestor. Man shares certain characteristics with several primates—the hands, feet and pelvis resemble the gorilla's; the skull and hair colour of the chimpanzee are close to man's; man has the same number of ribs as the orang utan, while his posture and gait is like that of the gibbon. Such pieces of evidence, only a few of many, indicate that in all probability the primates have arisen from a common stock.

Uakari *Cacajao calvus*
The uakari is a little known monkey found only in South America. There are two species, the bald uakari shown in this photograph, and its close relative, the black uakari, which is found in a separate range. The uakari is one of the smaller monkeys.

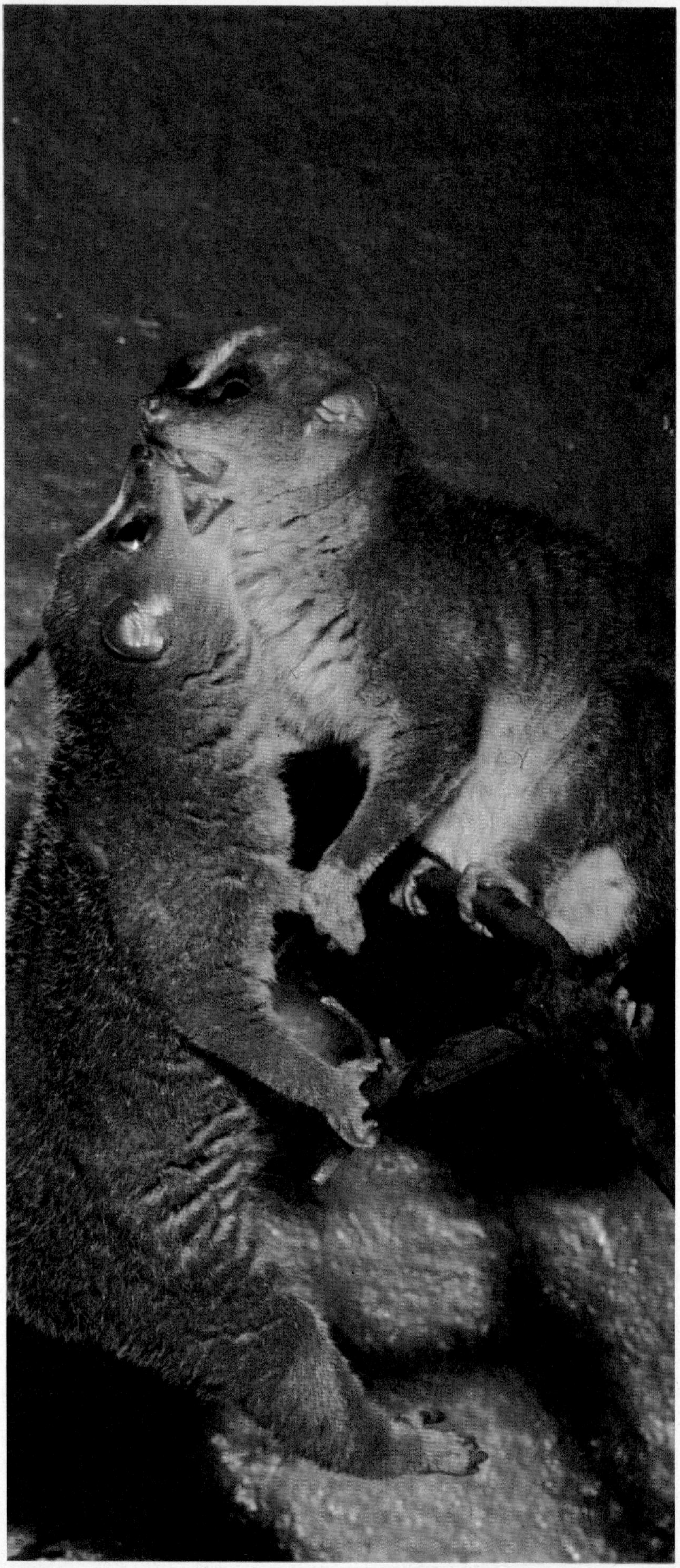

▲ **Gibbon** *Hycobates lar*
*This facial close-up of a lar gibbon gives a good view
of the prominent features and characteristics of the
primate face. There is a well-developed forehead and
the nose is reduced in size compared with most other
mammals. The jaw is thrust forward in the typical
manner, and for this reason primates are described as
prognathous. The eyes are the most notable feature,
both being directed forward, giving binocular or
overlapping vision which allows for accurate focus and
perception of depth for these animals.*

▶ **Angwantibo** *Arctocebus calabarensis*
*The angwantibo, a native of West Africa, is a member
of the lemur family, and is one of the most primitive
primates. This photograph of two angwantibos in
captivity shows several of their distinguishing
characteristics—the large, well-developed eyes typical
of animals which are nocturnal, and their highly
specialized feet. The first digit is opposable for
grasping, the second reduced to little more than a nub.
The remaining three are enclosed in a common web
of skin at their bases, giving the animal a powerful
caliper-like grip. As might be expected in a primitive
primate, the nose is not reduced in size.*

▶ ▶ **Orang Utan** *Pongo pygmaeus*
*Found only in Borneo and Sumatra, the orang utan is
one of man's closer relatives, and the great ape which
is in the most danger of extinction. Orangs have never
been numerous in recent centuries, and the gradual
destruction of their habitat and their low birth rate
have endangered their survival. Probably the greatest
damage has been done by collectors who capture
young orangs for zoos. The mother, who has only three
or four young during her entire lifetime, is usually
killed, and the survival rate of young in transit is low.
Consequently it is probable that every orang in a zoo
represents about 10 who have died in the process.*

▶ **Saki** *Pithecia monachus*

The sakis are a group of South American monkeys related to the uakari. They are found in the area south of the Amazon, near the Xingu reserve, and extend north to the Guianas. They are not particularly large animals. This photograph of a so-called monk saki shows the unusual features of this group. Like other North and South American monkeys, the nose is flattened and broad, with laterally-directed nostrils. The first digit is not opposable, but is just another finger or toe. The saki grips between its second and third digits. The most remarkable feature is the coat of fur which is long, rather luxurious and most unlike a monkey's. The long bushy tail appears to be quite out of keeping for a primate. The different colour of hair on the head and neck is also exhibited by other sakis, notably the white-faced saki and the white-nosed saki. Apparently the sakis dislike getting their fur wet, and in the wild drink by putting their hands in the water and then licking dry the dense fur on the backs of the hands.

The sakis live in the forest, near the edges or along river banks where the vegetation is particularly thick. Little is known about their habits. They are active tree-dwellers, constantly on the move. They are mainly fruit eaters, and except for moving to the shrubs and smaller trees to feed, spend all their time in the high canopy. They live in pairs and have territories which they defend vigorously. They are easily tamed, and make good pets.

▶ **Gibbon**

The gibbon is the smallest of all the apes. Their most distinguishing characteristic is their long arms, which may be 1½ times the length of their legs. Their hands are also remarkable in that the thumb has a very deep cleft, separating it from the palm. This allows the thumb to be kept well out of the way when the other fingers are being used for holding on to a branch, or swinging through the trees. The fingers terminate in claw-like nails.

All six species of gibbon live in southeast Asia and there is considerable colour variation, between the different species and also variation with age and between the sexes. Most gibbons are whitish when born, turning black, brown or grey during their first year. In three species, there is a colour change at sexual maturity, when the females turn to fawn but the males remain black.

The gibbon's acrobatic movement through the trees is spectacular. They swing along using their arms, apparently effortlessly, making huge leaps, gaining momentum by swinging their bodies. This ability is due to several factors. The body is lithe and light-weight and the well-developed eyes give binocular vision which allows the animal an acute perception of depth. The other notable feature of gibbon locomotion is its upright stance on the ground. Only man among the primates exhibits similar bipedal movement. Gibbons also use their legs for running along branches, balancing and steadying with their long arms.

◄ **Macaques** *Macaca mulatta*
The macaques are a group of 11 species of monkeys, and are the most widespread and numerous of all the group found in Africa, India, Tibet, Japan, southeast Asia and the neighbouring islands. They are generally small and brown, although a few species are black, and individual species are sometimes difficult to identify. Shown here is a macaque with a familiar name, the so-called rhesus monkey because of its association with scientific studies. The well-known RH factor, once a serious immunological problem in new-born infants and still occasionally a cause for concern, was first identified by medical researchers studying this species of monkey. A rhesus monkey was also the first primate to be shot into space in a rocket. Because of its scientific usefulness, this species is threatened in some areas, and tigher controls are necessary to force the institution of breeding programmes.

◀ Sifaka

There is some doubt among zoologists about the number of species and races of the sifaka. There are certainly many races, and there are probably one or two species. All the sifakas are found in Malagasy, where they are one of the most common lemurs. They live in trees at considerable heights and can readily leap from tree to tree. They can also swing through the branches with their arms, rather like the gibbons, and can walk upright on their hind legs, on branches or on the ground.

The sifakas are peaceful, relaxed animals. They live in small troops of about five animals. These troops are not family groups, but are associations of adults, although a single young may accompany its mother. The troops are highly territorial and territories are defended against neighbouring troops on the rare occasions that they meet.

They are strict vegetarians, the diet consisting mostly of fruit, with leaves, flowers and seeds making up the rest of the food. Sifakas do not appear to need to drink water, but presumably get all their moisture needs from their food.

They spend much of their time simply sitting in the sun and basking, and as they are highly social animals, also indulge in a considerable amount of mutual grooming. They have few enemies, principally harrier hawks, and fassas, which they try to frighten away by loud shouting and singing. The whole troop takes up the call in unison. If other troops in the neighbourhood hear this cry, they search the sky for predators and if they see any, they join in the call.

▲ Macaque *Macaca sinica*

One of the rarer macaques is the toque macaque a native of Sri Lanka. It is closely related to the rhesus monkey and occupies the same ecological niche. It derives its name from its little topknot which rather resembles a wig. Like all other species of macaque, it lives in troups which may be as small as 10 monkeys in poor terrain, but over 70 individuals in better environments. Each troup has its own home range which may overlap with the home ranges of adjacent troups. If two troups encounter each other during their daily movements around the range, the battle may last for up to 20 minutes, but there is rarely more damage than a few scratches and bruises.

◀ Loris *Loris tardigradus*

The lorises are another group of primitive primates, close relatives of the angwantibo. They are natives of India, Sri Lanka and southeast Asia. These small animals have large eyes characteristic of nocturnal mammals and the extended snout typical of a primitive primate. This photograph is of a young slender loris clinging to a branch and waiting for its mother to return. Lorises can spend long periods in this position. This ability is due to an intricate system of blood capillaries in the extremities of the fore and hindlegs which reduces blood flow and allows the animal to clamp securely on to a branch and remain immobile without tiring or becoming cramped.

One young is born to the pair, and it remains with the mother for a year. When angry or disturbed the loris emits a growl, chatters in a high-pitched tone and tries to bite. They are not easily tamed as pets.

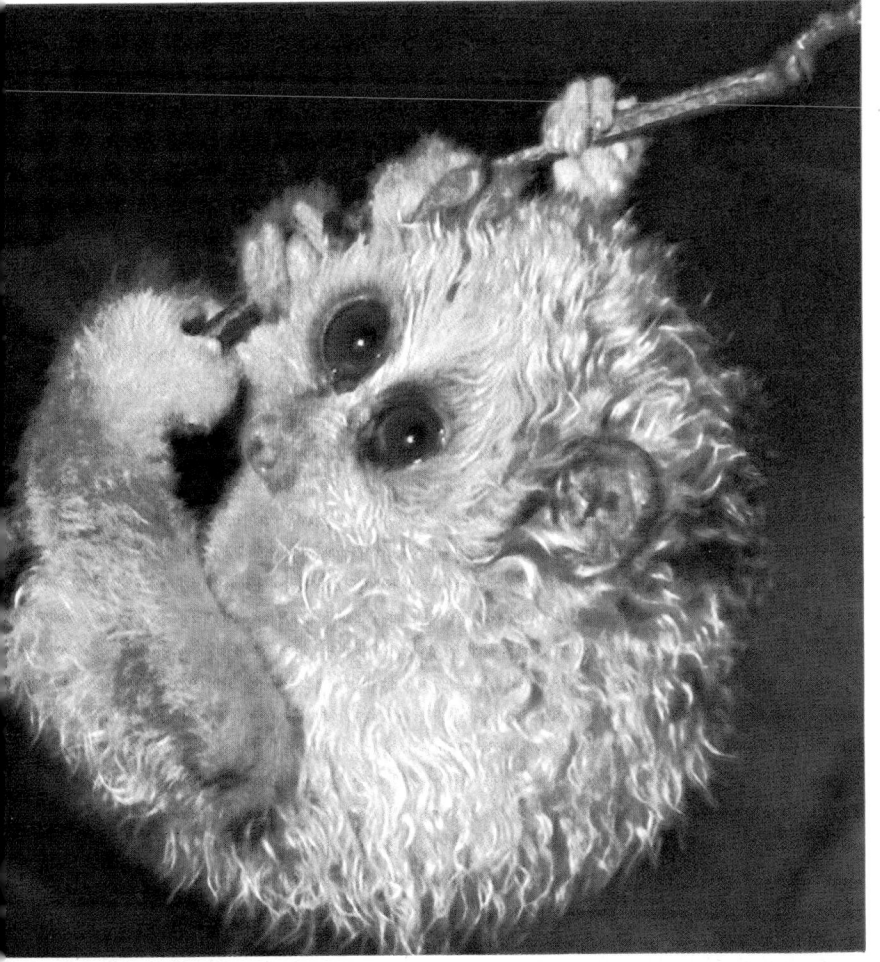

113

◄◄ Potto *Perodicticus potto*
A close relative of the loris is the potto, found only in
Africa in the forest belt which extends from Guinea
to East Africa. It is another primitive nocturnal
primate and shows all the primitive characteristics of
its relatives. Pottos are extremely slow-moving
animals, gripping the branch tightly with all four feet
and only releasing and moving one leg at a time. They
eat all manner of vegetation and fruit, as well as other
slow-moving animals—insects, snails, lizards and the
like. They will also raid nests for eggs and chicks, and
have even been known to catch birds. A curious
anatomical feature is the presence of four or five spines
on the neck, extensions of the spines of the cervical
vertebrae. It was once thought that these spines were
used as a defence mechanism, but as they are blunt and
in an odd position, this seems unlikely.

►▲ Langur *Presbytis geei*
The langurs are a group of about 20 species of
monkeys found in India, Tibet, parts of China,
southeast Asia, and the East Indies. Many of them are
brightly coloured, and some individuals may be of a
considerable weight. This is rare, however. Shown here
is a large golden langur which is particularly notable
in that it was only discovered and accurately
described as recently as 1955. It is a native of Assam,
and is found in the deciduous forests around the
Himalayan mountains. Two years later, in 1957,
another new species, the white-headed langur, was
described by a Chinese zoologist.

►▼ Barbary Ape *Macaca sylvanus*
The barbary ape is not an ape at all, but a monkey,
one of the many species of macaque. It is a native of
northeast Africa, and is famous as the ape which is
found on the Rock of Gibraltar, the only primate,
apart from man, still living in the wild in Europe. The
barbary ape is the only macaque to be found in Africa,
and was once widespread on that continent. Probably
environmental changes have cut them off from
their relatives in Asia, leaving this single isolated
species. They prefer a rocky environment with trees,
and spend their days roaming in large bands over the
landscape. At night they sleep in holes or in crevices in
the rocks. They appear to eat everything—most types
of vegetation, including items like pine cones, and
any small animals which they can catch. The colony in
Gibraltar is protected by the British Army.

►► Woolly Monkey *Lagothrix hendeei*
The derivation of this animal's name is obvious from
this photograph of a Hendu's woolly monkey. A native
of west central South America, there are only two
species of woolly monkey, the other being the little-
known Humboldt's woolly monkey, which is confined
to a small region between the Andes and the Pacific
Ocean. Woolly monkeys are quite large and have long
prehensile tails which they use to hang from branches,
leaving their hands and arms free for feeding on leaves
and fruit. It appears that they spend all their time in
the trees, jumping rather than swinging from tree to
tree. Open spaces are avoided by circling them or going
to a higher or lower level. Only in captivity have they
been observed on the ground.

◄ **Gibbon**
These two photographs illustrate colour variation in the gibbons. Above is a lar gibbon—*Hylobates lar*—a widespread species. It is a native of Borneo and Sumatra, and has a wide variety of colour phases. This photograph also illustrates the characteristic posture of a gibbon in 'flight'.

Below are two members of a family of siamang gibbons—*Symphalangus syndactylus*—the largest of all the gibbons. It has a restricted range in southern Malaysia.

Gibbons live in small family groups of two parents and up to four young of different ages. Only one young is born at a time, and it remains with its parents until about six years of age. Each family group has its own territory. Adults mate for life, and are extremely aggressive towards other gibbons. Each species of gibbon has a distinct call by which it identifies itself and recognizes others of the same species. Gibbons are omnivorous, although they seem to prefer readily available fruit like grapes.

► **Gorilla** *Gorilla gorilla*
The gorilla is the largest of the anthropoid apes. This photograph of a baby gorilla shows well its jet black skin and its grey-brown-black hair. The gorilla is found in a few isolated regions of central Africa, and can be divided into three distinct races. The gorilla has few natural enemies, except man, and is not in any way aggressive, although loud and fierce displays of anger when disturbed may give the impression of a highly aggressive animal.

Rodents

The rodents are by far the most numerous of all the mammals. They are distributed all over the world, and at least a few rodents are familiar to most people, especially those, like the common rat and the house mouse, which are closely associated with civilization. Not only are there more species of rodents than other mammals, but there may be more individuals as well.

The rodents are in many ways like the rabbits and hares. They are superficially similar, and all used to be classified as rodents by scientists, but it has now been established that the differences are sufficient to warrant the separation of the two, the rabbits and hares being placed in a group of their own, the lagomorphs. Similarities between the rodents and the lagomorphs are the result of parallel evolution rather than close relationship.

Externally, it is difficult to know which animals are rodents and which ones belong to another group. Examination of the teeth, however, provides definite evidence about the group to which an animal belongs. The rodents are gnawing animals. In addition to their high-crowned grinding molars at the back of the mouth, they have an upper and lower pair of chisel-like gnawing teeth, the incisors. These teeth wear away at the tip very quickly because of their heavy use but they grow out from the base just as quickly. The success of the rodents is a tribute to their way of life. They have adaptively radiated over most of the world, evolving special-izations which have allowed them to exploit a wide variety of ecological niches. In so doing, their external shape and size has varied greatly. Ground-dwelling rats, mice and chipmunks are very different from the burrowers such as gophers and ground hogs. The squirrels and porcupines have taken to a primarily arboreal way of life, while muskrats and beavers have become semi-aquatic.

Because of their size, numbers, and relative defencelessness, it is not surprising that the rodents form a large part of the diet of carnivorous mammals and birds of prey. They have survived this heavy predation by a combination of behavioural and physiological adaptations. Their senses, especially those of hearing and of smell, are extremely acute, giving warning of danger and the opportunity to escape. Many species are nocturnal, although of course, there are nocturnal predators as well, but relatively fewer than during the day. Most rodents have developed burrowing habits, or live in dense and concentrated cover.

Spiny Mouse *Acornys cahirinus*
The spiny mouse is a native of Africa and Asia. The typical species of the group is the Cairo spiny mouse seen here, so called because the first specimens of the species to be accurately described were from that city. The sharp prickles of the back and tail have evolved from soft hairs and are presumably a protection against predators.

120

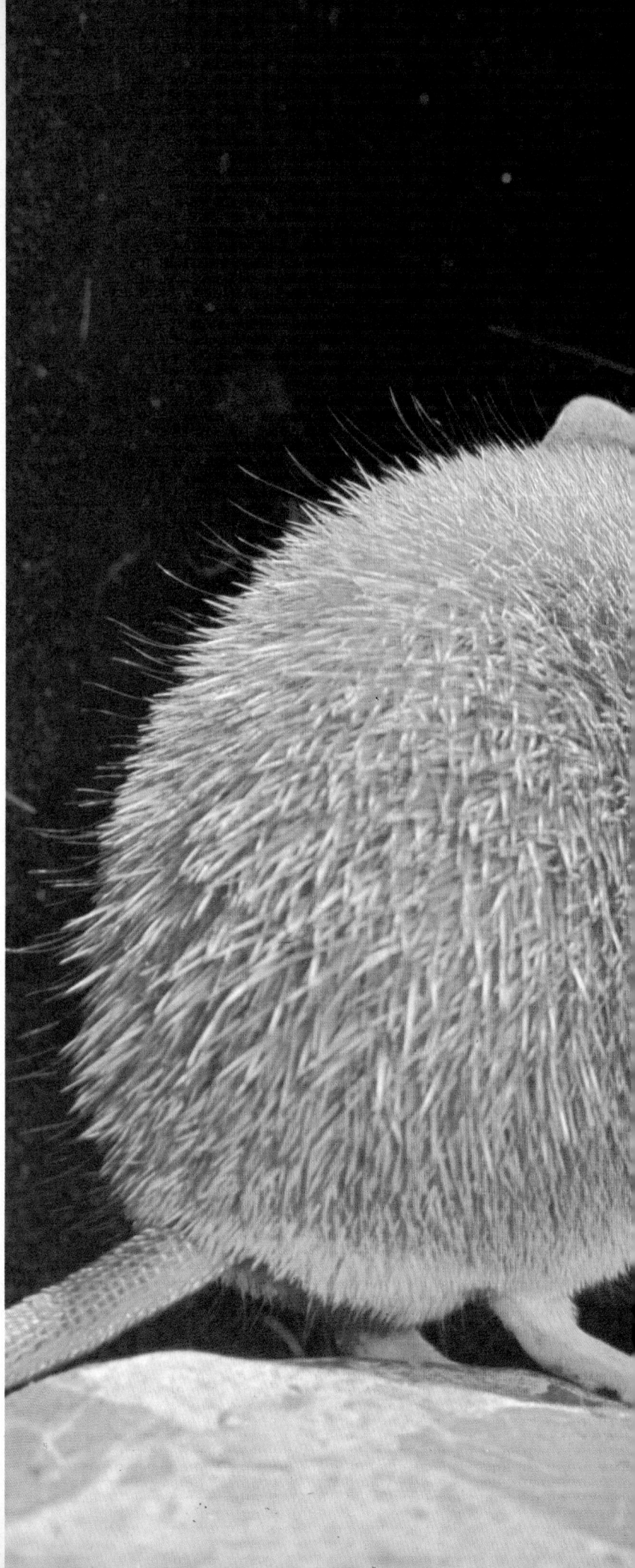

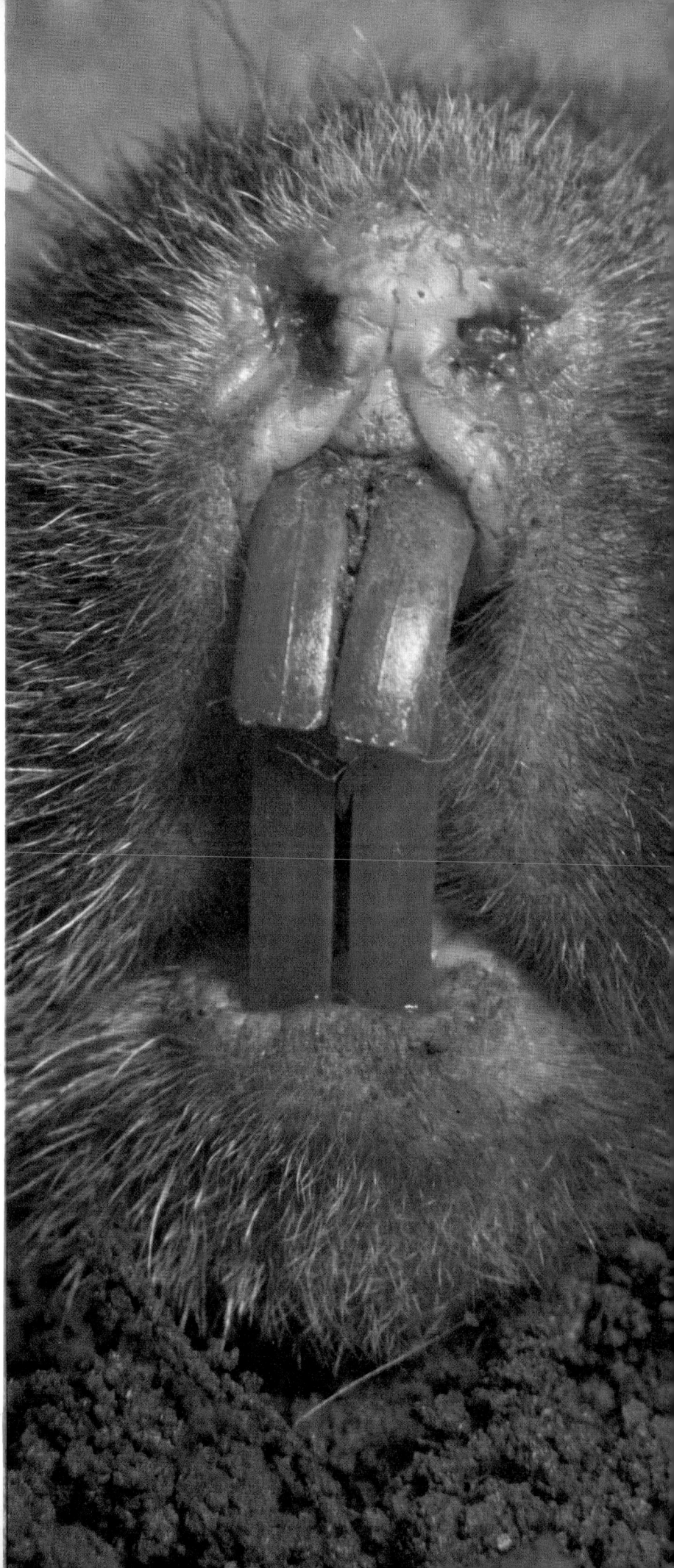

◄ **Viscacha** *Lagostomus maximus*
*There are three groups of viscachas, all of which are
exclusive to South America. The plains viscacha of
Brazil and Argentina is shown here. Its close relative,
the mountain viscacha, of which there are four species,
is found in a narrow belt in Peru, between the Andes
and the Pacific. A third species is now almost certainly
extinct. The viscachas are close relatives of the
chinchillas. They live in colonies of up to 30 animals,
and dig extensive inter-communicating burrows with
several entrances. Apparently some of these
communities have been in use for centuries. Each
colony is dominated by a single adult male. The
viscacha is a nocturnal animal, and on its nightly
rounds frequently visits the burrows of neighbouring
colonies. Burrows may be shared with other animals
such as snakes, lizards, ants and skunks, apparently
without conflict.*

   *The viscacha is a herbivore, eating leaves, roots
stems and seeds of most grassland plants. With the
decline of their primary natural enemy the puma, and
the increasing cultivation of the land, the plains
viscacha has become a serious economic pest,
damaging crops and grazing land and injuring
domestic animals which catch their feet in the
numerous openings to the burrows. Some areas of
Argentina have now been cleared of viscachas,
although eradication is a long, slow process, as
trapping appears to be the most efficient method.*

► **Mole-Rat**
*The mole-rats have derived their name from their
appearance and their behaviour, for although they look
more like moles, they are in fact rodents, and closely
related to the rat. Like many burrowing animals, they
are well-adapted to their underground life. They have
short dense fur which has no 'set', that is, it can be
brushed in any direction without appearing to be
irregular. One species, the naked mole-rat, has lost all
its fur except for a few sparse, long hairs. Their eyes
are reduced in size and sight is not good. The tail is
also similarly reduced or absent altogether. The
external ears are small or almost invisible. The legs
are short but powerful.*

   *Most species of mole-rats are found in Africa,
although one species, the Palestine mole-rat, is
widely distributed throughout the Middle East and
Russia. Unlike the true moles, who dig their tunnels
and burrows with their powerful forelegs and large
claws, the mole-rat makes great use of his over-sized
incisors. The jaw muscles of some species have
become so highly developed for this purpose that they
have filled in the orbits of the eye, while the eyes
themselves have become greatly reduced in size.*

   *Mole-rats eat the underground portions of most
plants, and can thus be a serious pest if there is an
infestation of them. They also store food in specially
prepared underground chambers against the time
when the ground may be too dry and hard for digging.
Their most common enemies are other digging and
burrowing mammals—badgers, foxes and jackals—
and owls and eagles, especially during times of
flooding or when the young are dispersing from the
home burrow.*

◀ **Harvest Mouse**

The female harvest mouse builds a small nest of grasses in which to have her young. The breeding season is from April to September, and five to nine young are born. The young become independent in about two weeks, so that the female can easily have several litters in a summer. Harvest mice eat a variety of cereals and grain, especially the seeds of grasses. They do not hibernate in the winter, but live in a burrow and continue to come out to feed. In the summer months, insects are included in their diet. Due to the increased use of farm machinery and changes in agriculture, they are now less common.

▼ **House Mouse** *Mus musculus*

The house mouse is doubtless the most common rodent. It is found throughout the world although it probably originated in Central Asia. Originally a native of the woods and fields, where many individuals still live, it is now most commonly associated with the artifacts and debris of civilization. It will eat any kind of food that it can get, and so is a considerable domestic problem. The white mouse, which is a common research animal and now a fairly common pet as well, is an albino strain of the house mouse and will breed with its wild relatives.

▶ **Harvest Mouse**
Micromys minutus
The harvest mouse ranges throughout Europe, except for the Mediterranean region, across northern Asia and China. Except for a few species of shrews, it is the smallest of the mammals. This photograph of an Old World harvest mouse shows some of its characteristic features. Because of an opposable digit on each hind foot, it is able to perch on suitable foliage and use its front paws for other tasks. The long tail is used not only for balancing, as here, but also for grasping. These features give the animal a remarkable degree of agility, and make many of its quick movements look almost acrobatic.

▶▶ **Gerbil** *Gerbillus pyramidum*
Because of its recent popularity as a pet, the gerbil is now a well-known rodent. Its natural habitat is the desert and the semi-arid regions of Africa, the Middle East and Asia. The gerbils are closely related to the common rat, although most species are much smaller. They have powerful, well-developed hind legs, which allow them to elude their pursuers with long leaps and some species can change direction with each jump.

Like most desert animals, the gerbil is primarily nocturnal thus avoiding the heat of the day. It is further adapted to its desert environment by being able to metabolize most of its needs for water from its food, and so it is independent of natural water supplies. The few diurnal species have dense fur and thick skin.

◄ Guinea Pig *Gavia porcellus*

The guinea pig, which is not a pig at all but a rodent, is a native of South America. There are three species of which one is the well-known domestic pet which has given rise to several varieties through selective breeding. As well as variation in the type of fur, guinea pigs have a wide range of colours and patterns, a few of which are shown in this picture of two adults and several young of different ages. Why they were originally confused with pigs is uncertain. They were first discovered in the Guianas, so no doubt the 'guinea' is a corruption of this. In spite of their timidity, they can make good pets, and their usefulness in medical research has given rise to the colloquial use of the phrase 'guinea pig'. In the wild, they live in small groups of from five to ten individuals.

▲ Deer Mouse *Peromyscus sp.*

The deer mouse is a small rodent native to North America. There are at least 55 species of deer mouse, and they range across almost the whole of the continent and extend into parts of Central and South America. No one species has a particularly large distribution. Most species are confined to particular areas, and although these may overlap, each one remains distinct in spite of their close relationships and their undoubted physiological capacity to interbreed. The species are probably kept distinct by behavioural factors, such as occupying different niches in the same environment, and by the recognition of behaviour patterns and smells. Each animal also has a home range to which it confines its activities and this probably contributes to the separation of the species.

▲ **Hamster** *Mesocricetus auratus*
There are 14 species of hamster, ranging across Europe, Asia and the Middle East. Best known of these is the golden hamster shown here, the species which has been domesticated and has become a fairly common pet. Surprisingly, hamsters have only been commonly known for less than 50 years, and it is thought that all the domestic hamsters are descended from one family group which was captured in Syria in 1930. Hamsters are also popular laboratory animals because they are clean and almost odourless, and they have yielded much useful information on aspects of cancer growth and control, hibernation and the immunology of tissue grafting. This photograph illustrates a distortion of the hamster's physiognomy. The animal has large, elastic cheek pouches, lined with tough skin, which it uses to store food, mainly plant material, before carrying it back to its nest.

▶ **Bank Vole** *Clethrionomys glareolus*
There are many species of voles. Like other small rodents, the voles are usually confused with their relatives and erroneously called 'mice'. Voles are specifically different from mice in that they have a snub nose, smaller eyes and less obvious ears, which are somewhat buried in the fur of the head. The bank vole is a common resident of Europe, extending east to central Siberia. In Britain, there are numerous subspecies mostly found on coastal islands. These are almost certainly descendants of the mainland species, introduced by man in prehistoric times. Over the centuries they have evolved separately from their ancestors, and have acquired enough distinguishing characteristics to be recognizably different from the bulk of the bank vole population. Thus there are several subspecies, each of which is specific to a particular island.

130

Cats

The cats are members of the group of animals known as carnivores, or meat-eating animals. Strictly speaking, carnivorous animals are those which eat any other form of animal life, but in general usage the term carnivore is usually reserved for those mammals which feed on other mammals. The rise of the carnivores was an almost inevitable result of the evolution of the mammals. As the mammals increased in numbers and in diversity of habitat and behaviour, the opportunity presented itself for some of them to feed on others, and certain members became specialized for a carnivorous way of life. The shift from a diet of insects was not particularly difficult. Two changes were necessary. The stabbing and shearing action of the teeth needed improving, and the structure of the legs and feet had to be adapted to allow the predator to run faster than his prey. The foot became longer, and the method of running became digitigrade, that is, running on the toes, but not the tips of the toes, rather than the more primitive plantigrade posture, in which the entire foot comes in contact with the surface over which the animal is moving. Several different groups of animals show these specializations, adapting thus for a carnivorous way of life: the weasels, dogs, racoons, bears and cats.

The cats are probably the best known and certainly the most easily identified group of carnivores. At one time, all cats were thought to be members of the same genus, Felis, and although this has now been modified, the relationships among the members of the cat family are still very close. For example, the lion, the domestic cat and the cougar are all members of the genus Felis, indicating a close relationship. Some cats are quite rare, but no-one who has ever had contact with any cat fails to recognize another member of the cat family.

One of the most interesting comparisons which can be made in the cat family is between those cats found in Africa and those native to South America. The leopard and the jaguar are very similar in form and habit, both species having a black colour phase. There is also a striking similarity between the African lion and the mountain lion. This latter cat was once widespread both in North and South America, having the largest natural distribution of any mammal but like so many other big cats, has been persecuted by man and forced out of his habitat by encroaching civilization. Many of the cats are now endangered species, or nearly so, and strenuous efforts must be made to ensure the survival of these magnificent creatures.

Lynx *Lynx lynx*
The lynx is a medium-sized cat of the northern hemisphere. Once common in northern Europe, it was almost exterminated as a pest, but with protection over the past few decades it is beginning to recover. Canadian lynxes were also decimated by the fur trade but are now protected. It is also widely found in northern Asia.

◄ Tiger *Panthera tigris*

The cats, big and small, show many identical or similar behavioural characteristics. Anyone who has ever watched a domestic cat move her young will immediately recognize that this female tiger is accomplishing the task in exactly the same way. The only difference would appear to be in the enormous difference in size between the house cat and this, the largest of all the big cats. Size combined with the tiger's distinctive, beautiful colouring makes it the most spectacular of all the cats and one of the most magnificent of all animals. Although a number of geographical races of tiger have been identified, varying in size, colour and markings, there is only one species. Tigers are found only in Asia, and specific populations are recognized in Siberia, the original home of the tiger, China, India, Iran, Java, Bali and Sumatra.

► ▲ Bobcat *Lynx rufus*

As will be obvious from its scientific name, the bobcat is a close relative of the lynx, although the two are easily distinguished by their markings and by the length of the tail. The bobcat's range is across most of North America south of Alaska and the Northern Territories to central Mexico. The bobcat, also known as the wildcat, seems to have coped with the spread of civilization better than most mammals of its type. This is probably due to its relatively small size, enabling it to hide easily, its wide range of food, its few natural predators and its ability to elude dogs and hunters by taking to swamps and to the water.

► ▼ Wild Cat *Felis sylvestris*

The wild cat, which is one of the European cats, should not be confused with either the North American wildcat, the bobcat, nor should it be assumed that it is descended from domestic cats gone wild. Further, there is no evidence to suggest that the domestic cat came from taming wild cats. The wild cat was originally found across most of southern Europe, extending east into Asia Minor. It is now confined to remote mountainous regions, and is thought to be on the increase in the Scottish Highlands. The wild cat is larger and more heavily built than the domestic cat. The tail is thick and bushy, although shorter, relative to body size, than that of many cats. The fur is long, thick and very soft. Coloration is highly variable, and it has been suggested that this is because there has been considerable interbreeding with domestic cats which have reverted to a wild existence. The decline of the wild cat is mostly attributed to man. It is regarded as a pest because of its destructiveness to domestic stock, primarily poultry and lambs, and has been actively hunted. Its survival has been due to the fact that most large European predators, its natural enemies, have now disappeared.

◀ **Leopard** *Panthera pardus*
The leopard, one of the biggest cats, is also one of the species whose survival is endangered, probably in general, and certainly in particular parts of its range. For many years, its beautiful skin has been prized in fashion circles, and although now there is considerable public pressure to stop this sort of exploitation, and there are strict controls on the hunting of leopards, there is still a lot of poaching and illegal exporting of skins.

Historically, the leopard has been regarded as a threat to domestic livestock, and although it has been shown that leopards do more good than harm in agricultural areas, there is still a large school of thought that portrays the leopard as a villain. The leopard is a native of Africa and southern Asia, concentrated in areas in which its favourite prey, primarily medium-sized herbivores, are located. Leopards will eat almost anything available and birds, rabbits, fish, frogs and insects are recorded as food sources. The panther is a black strain of the leopard, and close examination reveals an underlying pattern of the typical spots of the leopard.

▶ ▲ **Jaguar** *Panthera onca*
The jaguar, when seen in captivity is frequently confused with the leopard. The two are of about the same size, although the jaguar is more powerfully built, especially through the head, neck, shoulders, and forelegs. Its spots, unlike those of the leopard, are arranged in rosettes of four or five spots grouped around a single central spot. The jaguar is a native of the western hemisphere, where it is the ecological equivalent of the leopard. It is found throughout South America except for the southernmost parts and the high Andes, and ranges northward through Central America to the southwestern United States. Another similarity to the leopard is the not uncommon black variety of jaguar, which has also retained the underlying spotted pattern of its golden relatives. Because of its remoteness and its solitary existence not much is known about the jaguar in the wild, and much of our information about its behaviour is based on folk tales and anecdotes.

▶ ▼ **Tiger**
This photograph shows the inaccuracy of the tales that cats hate water. Certainly some cats, particularly the domestic cat, seem to avoid immersion in water, but not so the tiger. It is frequently seen in the water, and is a strong swimmer. In the rainy season or at other times when rivers are in flood, tigers have been seen to concentrate on fish and turtles as a food source. They also take to the water in search of stranded or swimming mammalian prey. Tigers dislike intense heat and have been found sitting in shallow bodies of open water, in the heat of the day, presumably in an attempt to keep cool. In spite of its aversion to hot climates, it has successfully invaded regions quite dissimilar to its original range. The tiger comes from the open snowy wastes of Siberia, but has adapted itself to mountainous regions, low lying areas and dense jungles. It prefers abundant cover and generally remains in the long grass or other vegetation during the day.

Genet *Genetta genetta*

The most important thing to note about the genet is that in spite of its appearance, especially its spotted coat, it is not a true cat, although it is frequently mistaken for or identified as one. It is most closely related to the mongoose, and lies in an intermediate group between the weasel family and the cat family. All six species of genet are found in Africa south of the Sahara desert. The feline genet, which is the most widespread, is shown here. It is found in all of southern Africa except the desert regions, as well as Morocco, Algeria and Tunisia, the Middle East, Iran, Spain and southern France. In contrast to this, the little-known water genet seems to be restricted to the area in the immediate vicinity of the east shore of Lake Victoria. Genets appear to be mostly solitary animals, coming together only for breeding, a brief period which occurs once and occasionally twice a year. Although it is a good mover on the ground, it is truly at home in the bushes and trees, where it is amazingly sure-footed and does most of its hunting. Observations on captive genets seem to indicate that each animal knows its own range so well that it can almost be said to have memorized it, and always follows the same route through it when hunting. This knowledge of the territory is so exact that the genet can move around in total darkness without making a mistake. Keen sight is doubtless primarily responsible, but touch and smell also play a large part in this remarkable self-taught process. They are typical carnivores with small but extremely sharp teeth. They eat any small animals they can catch including birds and insects.

◄ **Lion** *Panthera leo*
The lion is without doubt the best known of the non-domestic cats. Historically, it has been renowned for its strength and agility. Its relatively easy-going social nature, which makes observation in the wild a simple task, has meant that it has been well known by man for centuries. Lions were once relatively common animals. They ranged throughout southern Europe, almost all of Africa, the Middle East and India. They have now disappeared from almost all of that range. They are extinct in almost all parts of Africa except for the large game parks. A small intensely protected population of 200 or so in India is the only group found free outside Africa. Lions live in large family groups called prides of between 20 and 30 animals. There may be one or more adult males, a larger number of females, and a mixed group of juveniles and young. They are co-operative animals, combining their talents in hunting and defence. Females do the hunting and killing, but the males always eat first, followed by the females. Juveniles and young eat what is left, so in times of food shortages mortality among the newly-weaned may be very high. If food is abundant, then the females will hunt especially for their young. Lions may also let other animals such as hyenas, do their hunting for them, driving them away from their kill by brute strength and leaving the left-overs for the real owners.

▶ **Tiger**
The economic importance of the tiger's threat to man and man's environment has probably been grossly exaggerated. Certainly some tigers have turned to the killing of domestic animals, or have even become man eaters. However, these animals are exceptions, and are probably old animals who can no longer compete on equal terms, or ones which have been wounded and cannot hunt normally, and so have turned to easier prey. Evidence supporting this comes from a study which showed that tigers tend to disappear from an area when their wild prey disappears, even though there may be abundant domestic livestock to provide alternative food sources. Because of its size and strength, the tiger can prey on almost anything it chooses. Deer, antelope, wild pigs and occasional buffaloes make up a large part of the tiger's diet, but smaller animals, such as hares and monkeys are also eaten, and even swarming insects are considered edible. Adult tigers are solitary, coming together only for mating, which may be a period of only two or three weeks every two years or so. During the mating period, the male will defend the female to the death, but once she is pregnant, he deserts her. As many as six cubs may be born, but rarely more than two survive to become adults.

▲ Caracal *Felis caracal*
The caracal is also known as the caracal or desert lynx, and some zoologists believe that it is closely related to the lynx. It is a medium-sized cat, not unlike the lynx in some respects, most notably the prominent ear-tufts, thought to be of assistance to hearing. The caracal is found in Africa north of the Sahara desert, through the Middle East and Arabia to northern India, and also in east and south Africa. Its presence elsewhere—southern India and southwest and central Africa—is not certain, and it is now extremely rare even in some parts of its established range. Caracals feed on a wide range of small and medium-sized mammals and on birds. They have even been reported to be capable of catching birds in flight by leaping into the air. They also kill snakes, although probably not for food. They have a bad reputation for taking domestic animals, primarily poultry, and because they are superb leapers it is difficult to protect stock from them. Because of their mostly nocturnal habits, little is known about their behaviour or their domestic life. However, they have been tamed successfully and were once used in India for hunting.

▶ Serval *Felis serval*
The serval is another medium-sized cat, and is found only in Africa. It prefers open country with a relatively high rainfall, and may also frequent the edges of forests. Its range extends in a narrow band across west and central Africa to all of east and most of southern Africa. It has also been reported in north Africa, primarily Algeria, but these sightings have not been confirmed. In some ways the serval resembles a small cheetah. Its background colour is similar, although the size and pattern of its spots are different. It has long legs which are adapted for running and allow the serval to reach high speeds in pursuit of its prey. Its primary food source is rodents, especially the mole-rat, and it has been suggested that the serval's large ears are an adaptation which helps it hear its prey burrowing. Servals also eat birds and other tree-dwelling animals, for they are quick and skilful climbers, although they seem to prefer to spend most of their time on the ground. Lizards and insects are also killed and eaten. The serval is another animal whose behaviour in the wild is not well known. Apart from man, it has few enemies.

Other carnivores

In the introduction to the section on cats, the evolution of the carnivores was discussed, and a distinction made between their mode of running, on their toes, and that of their prey, who run on the soles of their feet. There is, of course, another group of carnivores who do not run at all. These are the so-called pinnipedes, or fin-footed carnivores, the seals and sea lions, but these will not be dealt with here. This section considers the weasels, dogs, raccoons.

These carnivores, like some of the cats, have been subjected to intensive pressure by man. All carnivores are fur-bearing mammals in the sense that they have relatively coarse, protective outer hairs and soft dense inner hairs. Thus, many of these animals have been favourites of the fur trade, along with several other non-carnivorous mammals. Others, especially the cats, have been prized for their colourfully patterned fur, which is stylish if not particularly warm. They have also been persecuted on economic grounds, although in almost every species of carnivore studied, the overall good they do for the environment, both directly and indirectly beneficial to man, far outweighs the damage that they cause.

However, the belief that carnivores are generally incompatible with the activities of man persists, and many species are still hunted to attempt to exterminate them. This has been particularly true of wolves and bears in the northern hemisphere. The actual damage caused by bears appears to be minimal. Wolves, which have generally been regarded as a menace, are slowly coming back into some degree of favour in large game reserves in North America, where they are instrumental in keeping down the numbers of deer which, because they are protected, are increasing too quickly for the land to support them.

One of the best examples of a much maligned carnivore is the hyena. It was long regarded as a scavenger, its unattractive appearance being taken as evidence of equally unattractive behaviour, but recent studies have shown that it is a misunderstood animal. Observations in the past have shown that hyenas tend to lurk in the vicinity of a feeding lion, waiting for the lion to finish his meal, then moving in to clear up what little is left. It now appears that quite the reverse is true. Hyenas, like many other dogs, hunt in packs. A lion, coming on a recent hyena kill, will drive off the hyenas, an easy job for a large animal.

Red Fox *Vulpes vulpes*
The red fox is a member of the dog family. It is found all across Europe and into central Asia, as far south as northern India; it is also found in northwest Africa. The north American red fox is a close relative but a different species. In spite of centuries of persecution, generally for ill-founded reasons, the red fox still thrives.

▼ Dingo *Canis familiaris dingo*
The scientific name of the dingo shows that it is a sub-species of the domestic dog. Its presence in Australia cannot be explained, for the mammals there are all marsupials. Dingo remains more than 6000 years old have been found in Australia, so it seems most likely that the dingo was introduced probably by the Aborigines, who may have brought pets from Asia. If so, then it seems probable that some of these would go wild. They hunt in family packs, and kill thousands of sheep and cattle every year. In spite of fencing, shooting, poisoning and large bounties, they still thrive.

▶ Stoat *Mustela erminea*
The stoat is a member of the weasel family, better known by its other common name, the ermine. It is one of the most widely-distributed mammals, being found across the northern hemisphere except for Iceland, central and southern Greenland and a few other Arctic Islands. This young stoat shows a mixture of summer and winter pelage. Six to nine young are born in spring and are raised by the female alone. After weaning, the young remain with their parents and hunt in a family party, which may join up with one or two other families. They tend to be nocturnal animals.

▶▶ Fennec *Fennecus zerda*
The fennec is the smallest of the foxes, a native of the deserts of north and central Africa and of the Arabian and Sinai peninsulas. Its most prominent characteristic is its over-sized ears. They are useful as organs for heat loss, a problem faced by most desert mammals who must devise ways of keeping their temperatures down without sweating, because of the need to conserve water. Another function is to detect sounds of prey, enemies or mates. Although it will drink water readily if it is available, the fennec can go for extended periods without it. Burrowing and nocturnal habits keep its temperatures down.

◄◄ Spectacled Bear
Tremarctos ornatus
This is the only South American bear. It is found in the foot-hills and the higher ranges of the Andes as far north as Panama. It derives its name from the white patterning on its face. This pattern is extremely variable and may be little more than a divided white patch, as in this photograph, or may form a large complete circle around the eyes.

◄ Raccoon *Procyon lotor*
There are seven species of raccoon, all of them found in the western hemisphere. The best known and most widely distributed is the North American raccoon shown here. It ranges from northern Canada to central America. Another species, the crab-eating raccoon, is widely distributed in the northern half of South America. The other five species are confined to islands. The raccoon has shown a remarkable ability to adapt to encroaching civilization. As their preferred habitat, the woods and brushland, have been destroyed, they have moved into open country. Their relatively unspecialized carnivorous diet has allowed them to exploit many food sources. They will also raid garbage dumps for refuse, and eat commercially prepared non-protein dog food and meal for grazing animals if they can get it. They can be a fairly serious economic pest, invading fields for ears of corn.

▲ Panda *Ailuropoda melanoleuca*
The panda, although generally supposed to be a member of the bear family, is in fact closely related to the raccoons. It was unknown to most of the world until 100 years ago, when it was described by a French missionary who had seen a skin of a panda in China. The panda is a very restricted animal, found only in Tibet and in the Chinese province of Szechwan. Little is known about the habits of this intriguing animal in the wild, because of its remoteness and geographical and political inaccessibility. Difficulties in getting it to breed in zoos outside its native China, and lack of information about its numbers in the wild have caused the panda to become an appropriate symbol for the conservation movement. The panda was formerly hunted by the Chinese and its decline was hastened further by the mishandling of the species by western collectors, but it is now increasing.

▲ **Mink** *Mustela vison*
The mink is a member of the weasel family. There are
two species of this commercially important animal,
the American mink which ranges over nearly all of
North America, and the rarer, more restricted
European mink. This never had a particularly wide
distribution, but its range is now almost completely
obscured by the North American mink, which has been
introduced to Europe. Mink have been used
commercially for more than a century, and it was only
a matter of time before the North American mink,
with its superior coat, was introduced on European
farms. Some escaped and breeding populations, which
can be a considerable nuisance, are now common.

▶ **Otter** *Lutra lutra*
There are many species of otter, and at least one
found on each continent, except Australia. The two
most widely distributed and populous species are the
Canadian otter and the European otter. The European
otter is shown here in a characteristic pose, about to
dive. Otters are among the most playful animals. No
matter what they are doing, they always seem to make
it into some sort of game, with diving, splashing,
chasing and sliding, playing a large part in the activity.
Presumably they are inactive when asleep, but it would
appear to be the only time. They are superb swimmers
as can be deduced from their shape, using their bodies
and powerful tails to push them through the water.

►▲ Coati

There are three species of coatis, found in separate geographical ranges in Central and South America. The species found in Mexico has been extending its range into the southern United States. Their social organization is particularly notable, in that the females, juveniles and young live in bands of about 20 animals, while the males are solitary, except in the breeding season. This behaviour was noted many years ago, and for a long time it was thought that there were two separate animals, the sociable coatis and the solitary coatimundis. Coatis hunt on the ground and in the trees, where they are almost as agile as squirrels. They eat many kinds of small invertebrates, as well as lizards and rodents.

◄ Marten

The marten's closest relatives are other carnivores such as badgers, ferrets and otters. It is not a particularly well-known animal, as it is solitary except during the breeding season, when family groups are formed, and because it is primarily nocturnal. Martens are found mainly in the temperate and sub-temperate zones of North America, Europe and Asia, although in the last they are found in some tropical areas. The photographs here are of a young European pine marten (above), and an American marten (below). The marten's thick silky coat has for many years been prized, as sable, by furriers, and has led to its extinction in some areas and decimation in others, particularly Russia and North America. However, it is now raised on farms in Europe, and has been vigorously protected in North America. Its numbers are now slowly increasing once again.

► Grey Fox *Urocyon cinereoargenteus*

The grey fox is a native of the Americas. Its range extends south from the Canadian border, through all of the United States except for the Great Plains, into Mexico, Central America and the northern tip of South America. This is a photograph of a young fox cub. The young are born in the spring, and there are usually three or four cubs in the litter. Both parents help to rear the young, which are weaned at about six weeks. At five months of age they leave their parents and lead an independent life. The grey fox is not well known, for it is usually solitary and nocturnal. It is particularly notable for being the member of the dog family which is most at home in trees, climbing easily and leaping from branch to branch and tree to tree.

Primitive mammals

There are two groups of primitive mammals, the monotremes and the marsupials. They both have the characteristics common to mammals, but they are much more primitive than most other mammals. They are generally natives of Australia and have not had to compete with the more advanced species. It is possible that the monotremes evolved earlier than and independently from the other mammals.

The marsupials are the pouched mammals. They are closely related to the higher mammals, but differ from them in their mode of reproduction. Young marsupials are born, after a very brief gestation period, in a very miniature state. However, their forelimbs, mouth and palate are relatively well developed. At birth they cling to the mother's fur and pull their way up her belly, following a 'track' which she has made by licking her fur just before birth. The young animal pulls itself to a pouch on the mother's belly called the marsupium, in which the nipples are located. Once in the pouch, the young immediately attaches itself to a nipple and remains there sometimes for several months, until development is complete and it is weaned. Even after weaning the young may return to the pouch in times of danger or inclement weather.

Today, most marsupials are restricted to Australia and the neighbouring islands. The opossum of North and Central America is one exception, and the opossum and opossum rat of South America are two more. It is not easy to explain this curious distribution. Fossil records indicate that the marsupials were once much more widely distributed, thriving in South America and present, if occupying an unimportant position, in North America, Asia, Africa and Europe. These four continents were probably connected, forming a sort of World Continent, while Australia and South America were island continents. Although both marsupials and higher, or placental mammals, co-existed for some time, at least in South America, the reconnection with North America brought with it an invasion of competitively superior placental mammals, and the marsupials were gradually eliminated, as they were in other parts of the world. Only in Australia, where they were cut off from the placental mammals, did they survive.

◄ **Anteater** *Tamandua tetradactyla*
Although anteaters feed almost exclusively on termites, ants and grubs, they are not true insectivores, but members of a group of South American toothless mammals. This photograph of a female carrying her young shows the typical feeding stance.

▼ **Pangolin**
The hair on the pangolin's back has become modified to form large overlapping scales. The rest of the body is covered with hair. They are tree climbers, and the long, strong tail acts as a fifth appendage for grasping or hanging. Pangolins are toothless.

◄ **Bat-Eared Fox** *Octocyon megalotis*
The bat-eared or big-eared fox looks more like a
jackal than a fox. It is a member of the dog family,
and obviously gets its name from its oversized ears.
Because of its habits, it seems unlikely that these
enormous ears have developed to give the animal
particularly sensitive hearing. It has been suggested
that they are an adaptation which increases surface
area and gives this tropical animal a more efficient
method of losing heat and reducing body temperature.
Bat-eared foxes have a primary diet of insects. They
seem to have a particular preference for termites, but
are known to eat small mammals, eggs and nestlings,
other small animals and even some plant material as
well.

► **Anteater**
The long probing snout has evolved to assist them in
their rather bizarre feeding habits. The front feet are
equipped with sharp claws for tearing apart termite
nests. When the nest is opened, the muzzle is inserted
and the long tongue, probes every nook, cranny and
passage. The tongue is covered with a mucilaginous
saliva to which the insects stick, and the prey is
stripped off inside the mouth when the tongue is
withdrawn from the nest. The size of the nose,
compared with the eyes and ears, seems to indicate
that the anteater has a highly developed sense of smell.
It moves along with its nose close to the ground,
probing cracks and holes. However, the long snout
may simply be for reaching into holes for ants.

Wombat *Phascolomis ursinus*

The wombat, a relative of the koala and the possum, looks a bit like a small bear but has habits more like those of a badger. It is a round, solid animal. There are two species of wombat, this one being a coursehaired wombat which is found in southeast Australia and Tasmania. The other species is the hairy-nosed wombat which inhabits the extreme south of Australia. It is slightly smaller and the fur is soft and silky and grey in colour.

Wombats are nocturnal and spend the day asleep in burrows which they dig out with their powerful claws. They are usually solitary animals, except in the breeding season. One young is born between May and July. It stays in the pouch to begin with, but even when it is old enough to run free, it remains with its mother, usually until it is about six months old. During this time she feeds it by tearing up grass and dropping the stems by the young animal so that it can eat the tender shoots. Apart from man, the wombat has very few enemies. They were ruthlessly destroyed by settlers because they tore down fences to reach crops or sheep pastures. Their burrows were also destroyed because horses tripped on them. They eat mainly grass and roots and sometimes fungi and the bark of trees. They can cause considerable damage to crops.

159

Kangaroo

The kangaroo is probably the best-known of all the Australian marsupials. There are 55 species of kangaroos, wallabies and wallaroos, of which five might be called true kangaroos. Two are illustrated here. The red kangaroo above is an animal of the open plains, living in herds or mobs of a dozen or so. This photograph illustrates particularly well some of the principal features of kangaroos: the relatively short and small forelegs, the huge and powerful hind legs and elongated feet and the thick muscular tail. On the right, a female great grey kangaroo stands upright while her young Joey grazes from the pouch. Young kangaroos are tiny and very immature when born. They make their own way to the pouch, climbing up the mother's abdomen by clinging to the fur. They begin to nurse as soon as they enter the pouch and can find a teat. A newborn kangaroo may weigh only a few grams and remains in the pouch for about eight months. After leaving the pouch, it remains with the mother and continues to suckle for at least another six months.

◄ **Platypus** *Ornithorhynchus anatinus*
The platypus is probably nature's greatest natural freak. It has a toothless bill which is duck-like, although soft and pliable, webbed feet, a beaver-like tail, and although a mammal, it lays eggs. The platypus lives in burrows on the banks of rivers of the eastern third of Australia. It is an active, nocturnal animal, and spends much of its time in the water foraging for crustaceans, worms and other small aquatic animals. The female platypus digs a long intricate burrow in which to lay her eggs. Usually two small eggs are laid, and they hatch in about a week or ten days. The young are very immature at hatching, and are four months old before they are weaned. Although the platypus is a mammal, it has no teats. Milk oozes through abdominal slits on to the fur, and the young lick it off. Once threatened by hunters who collected the platypus for its fur, it is now well protected and in no danger of extinction.

▼ **Flying Phalanger** *Petaurus australis*
The flying phalangers and sugar gliders, of which there are five species, are found in New Guinea as well as Australia. They behave in a manner similar to the flying squirrel, gliding from tree to tree supported by a 'parachute', a wide flexible membrane running from the front feet to the hind feet, providing a large surface area for supporting the animal in its gliding. When at rest, the membrane is folded under the body, as seen in this photograph of a sugar glider.

▲ Cuscus

The cuscus is another type of phalanger, and is often mistaken for a strange kind of monkey. It is a tree marsupial, and the 16 species of cuscus are found in New Guinea and Queensland. They are primarily nocturnal animals, and spend the day in thick foliage, clinging tightly to branches. They are wide-ranging feeders on both plant and animal life. Little else is known about them except that they breed continuously as the females always have at least one young in the pouch.

▶ Pademelon

The pademelon is a type of small wallaby. There are four species found along the coastal regions of New Guinea, eastern and southern Australia and Tasmania. Their original range was greater and they were much more numerous in the last century, but several factors have contributed to their decline. They were formerly hunted for food and skins. Bush fires and land clearing reduced their habitat or made it more suitable for larger marsupials who took the greater share of food resources. They are also slow breeders and cannot re-establish themselves quickly. Probably their greatest threat was the introduction of the rabbit which can out-compete them.

▼ Flying Phalanger

This photograph of a flying phalanger gliding shows why the membranes which permit it to do so are sometimes referred to as a parachute. The phalanger climbs to the top of a tree, chooses a suitable target, and leaps into space, spreading its parachute as it leaps. By rotating the body, using the tail as a rudder, it is able to exert considerable control on its direction and altitude. It comes in for a landing on the target tree by meeting the tree flat on with all four feet at once, hanging on by its long sharp claws. Considerable altitude is obviously lost during this manoeuvre, so it is necessary for the phalanger to climb to the top of the tree before launching itself again. In spite of their accuracy in reaching their targets, phalangers sometimes land on the ground, probably by mistake. They are awkward runners—the parachute gets in the way—and quickly make their way to the nearest tree. Phalangers eat a variety of insects and small mammals, as well as birds, flowers, fruits, nectar and sap from plants. They seem to have few natural enemies, probably only the red fox on the ground and the so-called powerful owl in the trees. Probably the phalanger's greatest enemy is the continuing destruction of its traditional home, the eucalyptus tree. Not much is known of their breeding habits. Two young are born and these are carried in the mother's pouch until big enough to ride on her back.

▶ Rock Wallaby *Peradorcas xanthopus*

The rock wallaby takes its name from the rocky habitat in which it lives. Like all kangaroos, it is a leaper, and is especially sure-footed. The hind feet are provided with thick pads with rough surfaces, a built-in non-skid device. All locomotion is accomplished by the hind feet, with the tail being used as a balancing organ. Their agility in their rocky environment is prodigious. They can leap easily across chasms, scale leaning trees, pass through openings not much wider than themselves without interrupting their rhythm and climb what appear to be vertical rock faces which have no possible footholds. There are several species of rock wallabies some of them hare-sized, others rather larger. The one pictured here, the ringtailed wallaby is one of the most colourful kangaroos, with yellow markings on the face, limbs and feet, and alternate yellow and dark incomplete rings on the tail. They are found in all parts of Australia. Some writers have reported that the rock wallaby is nocturnal, but this may only be in regions in which it is hunted, or in which there are many predators. They appear to like to sit out on the rocks and sun themselves, some species even withstand the high temperatures of midday. They are vegetarians, and can go for long periods without water, apparently having the ability to extract the maximum liquid from their food. When the grass has dried out, they eat bark and leaves.

Brush-Tail Opossum *Trichosurus vulpecula*
*The brush-tail, vulpine or fox-like opossum is the
most common and most widely distributed Australian
marsupial. It is about the size of a red fox, and
has a fox-like head with prominent eyes and ears and
a long pointed snout. Its other most obvious feature
is its large bushy tail, which is prehensile. The
fur is thick and woolly and in the past these
animals have been hunted for their fur, in some years
more than a million specimens being taken. In spite
of this, and no shortage of natural predators, the
brush-tail thrives. This is probably due to two
factors. It has a preferred food, mistletoe, but will
readily eat many other kinds of food if this is not
available. It also adapts readily to new
environments, and is found in suburban areas, living
in buildings under the roofs. The brush-tail was
introduced to New Zealand in the nineteenth century
and the population there is now estimated to be 25
million.*

▶ ▶ **Koala** *Phascolarctos cinereus*
*This popular bear-like animal is often thought of as
even more Australian than the kangaroo. Unfortunately,
it is one of the world's most threatened species.
They once numbered many millions, but years of
shooting them for sport, hunting them for fur and
reducing their habitat, for they eat only eucalyptus,
have taken their toll. Rigorous efforts are now
being made to protect them and their environment and,
their numbers seem to be on the increase. They spend
most of their time in the trees, occasionally
climbing down to lick earth as a digestive aid.*

Flightless birds

All the flightless birds, with the exception of the penguin, are often grouped under the general heading of 'ratites'. In fact, the ratites are so grouped for convenience, for there are seven quite distinct unrelated groups under this term. Living examples of flightless birds include the ostrich, kiwi, rhea, emu and cassowary. Several species, including the dodo, have become extinct due to man's interference, some, such as the giant moa and the elephant bird, within living memory. All of the ratites have powerful legs for running, small wings, and lack a keel on the breastbone. All these characteristics are compatible with their way of life and mode of locomotion.

There is no doubt that the groups of flightless birds evolved separately from species which could fly. Their similarities are explained by convergent evolution, which means that animals which have a similar mode of life develop similar physical characteristics in spite of their quite separate origins. Convergent evolution is particularly well seen in the fish-like shapes developed by the large swimming mammals. In a way, the evolution of the flightless birds has been a sort of backwards evolution, or reversion to type. The wings, which are useless, have become small and rudimentary. The bones, hollow and lightweight in flying birds, have become solid to withstand the impacts inevitable in a ground-dwelling existence. The legs have become long and muscular for running. The keel on the breast-bone, to which the flight muscles are attached in other birds, is reduced or absent.

Ostrich *Struthio camelus*
*The ostrich is the largest living bird. Male
ostriches are polygamous. At the start of the
breeding season they establish territories away from
their communal feeding grounds, and the females join
them there. Each male has three to five females in
his harem. The nest is a simple affair, a depression
in the ground, and all the females lay their eggs in
the one nest. When egg laying is completed, the
dominant female drives the others away, and remains
to guard the nest with the male. Rather than
incubating the eggs, the female's job is to protect
them from the hot sun by shading them. Incubation
lasts about six weeks, and when the chicks hatch
they are able to fend for themselves in a
rudimentary way. The family group stays together for
a time, and then joins up with other families to
form large bands. Ostriches do not breed until they
are at least four years old.*

◀◀ Trumpeter *Psophia crepitans*
The trumpeter is not truly flightless. It can fly, poorly, if absolutely necessary, but prefers to move about on the ground and escape from predators by running away. It will even swim across water in preference to flying over. There are three species of which this, the common trumpeter, is the most numerous and widespread. They are all found in the Amazon.

◀ Ostrich
If the ostrich nest is threatened, the adults will indulge in vigorous defensive display. Here a female takes an aggressive stance to frighten away the invader. It is a long way from the 'head in the ground' myth. This sort of physical display is typical of many birds, and other animals as well. In this case, the wings and feathers are extended as far as possible.

▼ Rhea
The rhea is the largest flightless bird of the western hemisphere. The two species are native to South America. They move about in large flocks, splitting up only in the breeding season. The males are polygamous and the females polyandrous, so the parentage of young rheas is impossible to establish. The eggs are laid in large communal nests and the males alone incubate them.

Ungulates

The ungulates are a scientifically artificial grouping of the mammals. The name refers to their foot posture, which is said to be unguligrade, which means that only the hoof is in contact with the ground. Ungulates include animals like horses and cattle which have become specialized for a plant diet. Considerable changes in dentition, including the provision of a much greater surface area for the grinding of their food, have facilitated the adoption of the herbivorous way of life.

Not surprisingly, herbivores constitute the primary food source of carnivores. Ungulates protect themselves by the simple expedient of running away, and this has been made possible by the development of a foot well adapted for speedy running. The foot has

been lengthened and the animal moves about on the tip or tips of its toes. One or two toes reach the ground and the others have become vestigial or have disappeared. The primitive claw has become transformed into a hoof. The various contempory ungulates can be divided into two groups.

In the perissodactyls, the axis of the foot passes through the third toe, which is always the largest. A few of these animals such as the tapir and rhinoceros, retain the middle three toes, but in most, such as the horses, only the middle one remains. These are therefore the odd-toed ungulates.

The artiodactyls are the even-toed ungulates. In them, the axis of the foot passes between the third and fourth toes, which are of equal size and import-ance. Some artiodactyls, like the pigs, have retained four toes although the outer two are reduced in size. In others, such as deer, vestiges remain, but in most animals, of this group—giraffes, sheep, cattle and antelope—only the third and fourth remain. In all probability the perissodactyls and the artio-dactyls evolved separately, and owe their similarities to parallel evolution rather than to a common ancestry.

The artiodactyls are divisible into two further groups on the basis of their eating habits. The ruminants, or cud-chewers—cattle, sheep, goats, antelopes, deer, giraffes and others—regurgitate their food back into the mouth for further chewing. The non-ruminants—pigs, hippos and such—do not.

◄◄ **Springbok** *Antidorcas marsupiacis*
*The springbok is one of the species of African gazelle.
There is only the one species, but at least two
subspecies, are recognized, one confined to Angola.
There are also several relatively rare colour variations
in certain populations. This is probably the result of
the once enormous herds being broken up into smaller
populations by the reduction of numbers and the
fencing or destruction of their habitat, and by the
domestication of some animals. In smaller
populations, cut off from others, natural genetic
variations are more likely to be reinforced because of
the higher rate of inbreeding, rather than disappearing
quickly, as they would do in great herds. The
springbok gets its name from its habit of jumping.*

▲ **Rhinoceros** *Diceros bicornis*
*The rhinoceros is a vanishing animal. Because of its
prehistoric appearance, it is often assumed that it is an
evolutionary failure, and is disappearing for this
reason. This is not true. It is disappearing because it is
being systematically exterminated by man. There are
five species of rhino and despite their superficial
similarities the three Asian species are quite different
in many ways from the two African species. Shown
here is one of the African rhinos, the black rhinoceros.
The other African species, much more restricted and in
danger of extinction in the near future, is the white
rhinoceros. The three Asian species, the Indian, Javan
and Sumatran are similarly threatened, the Javan
species now existing only in one small reserve.*

▶ ▶ Bontebok *Damaliscus pygargus*
The bontebok is a member of the antelope family, and a native of the coastal plains of South Africa. It was never a particularly numerous animal, and as it occupied one of the first areas of South Africa to be settled, it was an immediate food source for the settlers. As agriculture developed, the bontebok were hunted even more, as they tended to feed on crops and in spite of the recognized danger of extinction, heavy fines for killing them and licences to shoot them, their numbers continued to decline. They were preserved only by the efforts of a few farmers in the last century, who decided to enclose them to keep them off agricultural land and to protect them. The only remaining bontebok, descendants of these survivors,

are found only in such reserves. Their close relatives the blesbok, which may even be the same species in spite of different outward appearances, were natives of the highlands further inland, and escaped major pressure on their numbers. The two animals occupy almost exactly the same ecological niche, but as they do so in completely different ranges, they do not compete for food sources. Their only defence from predators is by running. They are strong, fast runners, and have considerable endurance. Young bontebok develop muscular co-ordination quickly, and can outrun horses when only a week old. The blesbok and the bontebok interbreed readily in captivity, but the offspring have not yet proved fertile. This suggests that they are different species.

▲ Reedbuck

There are three species of reedbucks, one of the African antelopes. They are found in large overlapping ranges in most areas of open grassland south of the Sahara desert. They are defenceless except for the ability to run from pursuers, but even so they are not particularly fast, nor do they seem to have much endurance. The only defence of the young is to lie motionless in the long grass to avoid discovery. There is a wide variety of predators—most of the large carnivores, including lions and leopards as well as hyenas, jackals and wild dogs. Eagles and even pythons are known to prey on reedbuck. They live in small herds of up to 20, and probably the secret of their survival is connected with their numbers.

▶ Roan Antelope *Hippotragus equinus*

This African plains dweller is found over a wide range from South Africa to Ethiopia and Senegal. They are animals of the open woodland, browsers rather than grazers. They live in small groups or in pairs, and some, especially males, are solitary. They may often be found running with herds of gnu or zebra. Unlike other antelope, the roan has a reputation as a fierce fighter and will vigorously defend itself when attacked. Lions may take a few young or old animals, but they are otherwise more or less undisturbed by other predators except for man. This photograph also includes two oxpeckers hanging from the roan's head to drink. Oxpeckers are familiar social parasites of large grazing animals, feeding on lice and ticks.

▲ **Mule**

The mule is one of those rare animals which is produced by the interbreeding of two quite separate species. The mule is the offspring of a mating between a female horse and a male ass. Although viable in almost every way, it is itself infertile and cannot breed. Obviously, the horse and the ass are close relatives, both are members of the genus Equus, and their behavioural and other traits have not diverged sufficiently to prevent interbreeding. The reverse cross—the offspring of a female ass and a male horse is the less well-known hinney. It is a small horse-like animal, also sterile and relatively uncommon. This is probably because it is economically unimportant.

◄ **Sable Antelope** *Hippotragus niger*
*The sable antelope is a close relative of the roan
antelope. The two share much the same range in the
south, but the sable extends north only as far as
Angola and Kenya. They live in herds of up to 40
animals of varying ages, dominated by a single male.
Young males leave the herd as soon as they are
sexually mature, and remain solitary until they
acquire a herd to themselves. They live in light
bushland, and graze on the open country, thus
avoiding competition with the roan antelope. Like
the roan, they are fierce and courageous fighters,
and thus have few natural enemies. When a herd of
sable antelope moves to a waterhole to drink, other
animals make way for them. Because of destruction of
their habitat and indiscriminate hunting by man, their
distribution is now discontinuous, and some
populations, especially the race known as the giant
sable antelope, are in danger of extinction. Already
one close relative, the blue-buck, has been
exterminated. The blue-buck was an antelope
restricted to a particularly small part of southern
Africa, and was first described by a scientist in the
mid eighteenth century. By the end of that century it
was extinct, wiped out by hunting parties of Boers who
shot as many as they could whether they needed them
or not, leaving thousands of wounded animals to die
on the plains. Other species and races suffered the
same fate, and many of those that have survived have
done so only through determined conservation.*

▶ **Giraffe** *Giraffa camelopardus*
*The giraffe is a native of Africa. At one time it had a
wide range, which probably included North Africa
and Egypt, but it has been extinct there, probably for
about 2000 years or more. Its range is now confined
to a number of isolated semi-arid regions south of the
Sahara desert. Its distribution was once more
continuous than this, but slaughter for hides and
encroaching civilization have taken their toll.
Physically, giraffes from different regions vary
considerably, particularly in height, colour of coat and
pattern of spotting, but these are thought to be minor
variations in the one species, rather than an
indication of several species. Eight to thirteen races of
giraffe have been identified by different zoologists.
Giraffes are most notable for being the world's tallest
animal, although females do not reach the great height
of mature males. This enormous height allows the
giraffe to browse on high foliage, assisted by
particularly mobile, extensible lips and a long
powerful tongue. The preferred food is said to be
acacia, but this probably varies from region to region,
depending on what is available. Giraffes live in
sexually segregated herds, the males keeping to the
forested areas, while some older males may be
solitary. They join the herds of females, juveniles and
young only for mating. Young giraffes are as tall as a
man at birth, and can walk and run when only an
hour old. They stay with the mother for some time, but
the bond is not a strong one, and the young may be
more or less independent within a week or so.*

◀ **Dik-dik**

The dik-dik is the collective name for a group of six species of small antelope found in Africa south of the Sahara desert. It is a shy, elusive animal, and because of its small size and weight, hides easily in the long grass and brush. It is a mainly solitary animal, although pairs and groups of three, possibly family groups, are not uncommon. They take their name from their call, which they utter when disturbed or frightened. Because they are so timid and easily startled, they are the bane of the hunter's existence, warning other game of the presence of predators.

▼ **Oryx** *Oryx dammar*

The oryx, with its soft coloration and graceful recurved horns, is one of the most beautiful antelopes. There are three species of oryx, one in Arabia and now in danger of extinction, and two in Africa. All live in desert areas. This is a photograph of the scimitar oryx, a native of the Sahara desert. The closest relatives of the oryx are the sable and roan antelopes with whom they share many of the same characteristics. Physically, except for their horns, they are very similar. The oryx is particularly well adapted to its hot desert environment. As the temperature of the air rises, so does the body temperature so that a considerable increase is possible before the animal begins to sweat, the physiological adaptation which helps to cool the body. The oryx can also live for long periods without drinking. It feeds by night on desert plants, which are also adapted to their arid environment by being able to absorb water from the air at night, when the humidity rises. Thus the oryx, by eating at a time when its food has the maximum moisture content, is able to derive most of its water from its food. Oryx usually live in small herds of about 12 individuals, although as many as 60 in a herd have been recorded. The males indulge in display and fierce fighting before mating.

▶▶ **Red Deer** *Cervus elaphus*

The red deer is such a common name for deer that it is always difficult to know which particular species is under consideration. This is the European red deer, but it has close relatives, all members of the genus Cervus, in Asia, north Africa and North America, where the name red deer is also applied to the local population. In this photograph, two males are engaged in combat. This is a common sight in autumn as the usually solitary males round up harems of females at the beginning of the rut. The male constantly patrols the perimeter of his territory in which the females are kept, fending off any male who comes too close or who attempts to break up the harem. The battle that breaks out consists of ramming and butting in head-on clashes. The antlers which distinguish male red deer and their relatives are used extensively in conflicts, but there is some doubt as to the actual need for antlers. It has been suggested that antlers are signs of sexual superiority and primarily for defence, but more recent evidence indicates that their primary function may be for heat radiation during the summer months, when heat loss is necessary, while their signalling and defensive functions are very much secondary uses of these spectacular appendages. The stags cast their antlers between February and April.

▼ Pronghorn *Antilocapra americana*
*As its scientific name suggests, the pronghorn is a
native of America. It is found in the eastern half of
North America, from Mexico to the western
provinces of Canada, in rocky deserts and the
grasslands. It is frequently called the pronghorn
antelope, but it is not a member of the antelope family.
It takes its name from the distinctive shape of its
horns, which is well illustrated in this photograph of a
male pronghorn. Both sexes grow horns, and like deer,
shed these every year at the end of the breeding
season, growing a new pair in the next year. Once
the pronghorn population was numbered in millions,
but by the early part of this century had been reduced
to a few thousand. They are now protected, and again
on the increase.*

▶ Cattle *Bos taurus*
*There are two species of cattle, the western cattle
shown here, and the zebu of Asia and Africa. The
western cattle were derived from the aurochs, wild
animals which lived in the forests of Europe and which
became extinct in the seventeenth century. Although
they are often described as wild cattle, there is really
no such animal, for the name cattle carries with it the
distinction of domestication. When domestication first
occurred is not certain, but it was probably at least
5000 years ago. The almost infinite variety of cattle
testifies to the length of time they have been
domesticated and the efforts of man to produce
different races with particularly profitable
characteristics. Cattle were easily domesticated
because they naturally live in herds.*

Other mammals

This final section, includes several animals which mainly belong to groups not considered elsewhere in this volume but which are too important or too interesting to omit.

The warthog and the hippopotamus are both members of the artiodactyl group of mammals. These are even-toed ungulates, and the pig family, to which both these animals belong, is notable for having all four toes present rather than a reduction to two. The second and fifth, although reduced in size, are still functional, and it is thought that they have been retained because of the soft ground habitat which the pigs prefer. The wider foot gives the animal more support and mobility in a wet or mushy soil. The pigs and hippos belong to the non-ruminant sub-division of artiodactyls.

The elephants are plant-eating animals which are sometimes referred to as sub-ungulates, because they show certain ungulate characteristics in a less well-developed form. They walk on the extended tips of their toes, but all five toes are present, and a large pad of elastic tissue between the digits and the ankle is primarily responsible for bearing most of the body weight.

Elephants are close to being the maximum size possible for any land animal. It is impossible to see how a much larger animal could be 'engineered' so that its various systems could function in such a massive amount of living tissue. Large mammals have a relatively lower metabolic rate than small mammals, because the surface areas, through which heat is lost, are proportionately greater in a small animal than a large one. However, elephants still need large quantities of food. Food gathering has been made much more efficient for these animals by the development of the trunk, which is in fact a tremendous elongation of the upper lip and nose. By using the trunk, the elephant can gather food in bulk from a range of levels, and can drink without leaning over by filling the trunk with water and then squirting it into the mouth. The trunk gives the elephant family its scientific name, the proboscids.

Elephants also have a unique dentition. All the front teeth are missing except for one pair of incisors which have become modified to form the enormous tusks. The pre-molars and molars, which are massive, appear one pair at a time.

Warthog *Phacochoerus aethiopicus*
Not exactly the loveliest of mammals, the warthog is further handicapped by its rather unattractive name. It is a member of the pig family, and is found in open country in Africa from Senegal and Ethiopia in the north to South Africa. Its chief predator is the lion, but it is also hunted by other large carnivores, including man, against all of which it is apparently fearless, defending itself vigorously with its long tusks.

Elephant *Loyodonta africana*
These are two species of elephant, the African shown in these two photographs, and the Indian elephant. The African elephant is the largest living land mammal, found in most parts of Africa south of the Sahara desert. The smaller Indian elephant, or more properly Asian elephant, is found in the dense forests of southern Asia. The elephant is a herbivore, feeding on grass, foliage, branches of trees and fruit, which it grasps and pulls loose with its muscular trunk. Overpopulation of elephant, which has occurred in some areas where they are well-protected, can result in severe damage to the environment and the subsequent starvation of many members of the population. This has necessitated culling.

▶ **Hippopotamus** *Hippopotamus amphibius*
This distant relative of the pig is found in Africa. There are two species, the pigmy hippo, found in two restricted areas of west Africa, and the more common hippo shown here, found in the rivers of central and southern Africa. Its range was once much larger, but it has now unhappily become extinct in many areas.

▶ ▶ **Sloth** *Bradypus tridactycus*
These South American mammals are relatives of the anteaters and armadillos. They are well adapted to their aboreal upside-down existence shown by this female three-toed sloth and her young. Their long curved claws function like hooks. The arms are longer than the legs and the back and pelvis muscles are small and poorly developed.

Camel *Camelus dromedarius*

There are two species of camel, the one-humped or Arabian, and the two-humped or Bactrian. The Arabian is a domesticated version of the Bactrian, presumably developed by selective breeding. Given the opportunity, the two species readily interbreed. The one-humped camel, which is shown in these three photographs, is often wrongly called the dromedary, perhaps because of its scientific name. In fact, the dromedary is only one of several breeds of Arabian camel, one specially developed for riding. There are many stories about the camel's ability to go for long periods without water, and as many explanations for this ability, but most of these have been misleading, and many quite false. Camels can live for about a week or more without water, but must have wet vegetation to keep going. They cannot store water in their stomachs, nor can they convert the fat in their humps into water. There are two principal adaptations which allow camels to survive in the desert. Their long legs and necks provide large surface areas for heat loss, although water is also lost in this way. They also have a remarkable physiological adaptation by which water is withdrawn from the tissues of the body but not the blood. Because of this, the blood does not thicken with water loss from the body, as it does in other animals, causing additional strain on the heart and circulatory system. A camel which has gone without water for some time may drink 100 litres of water or more in a few minutes.

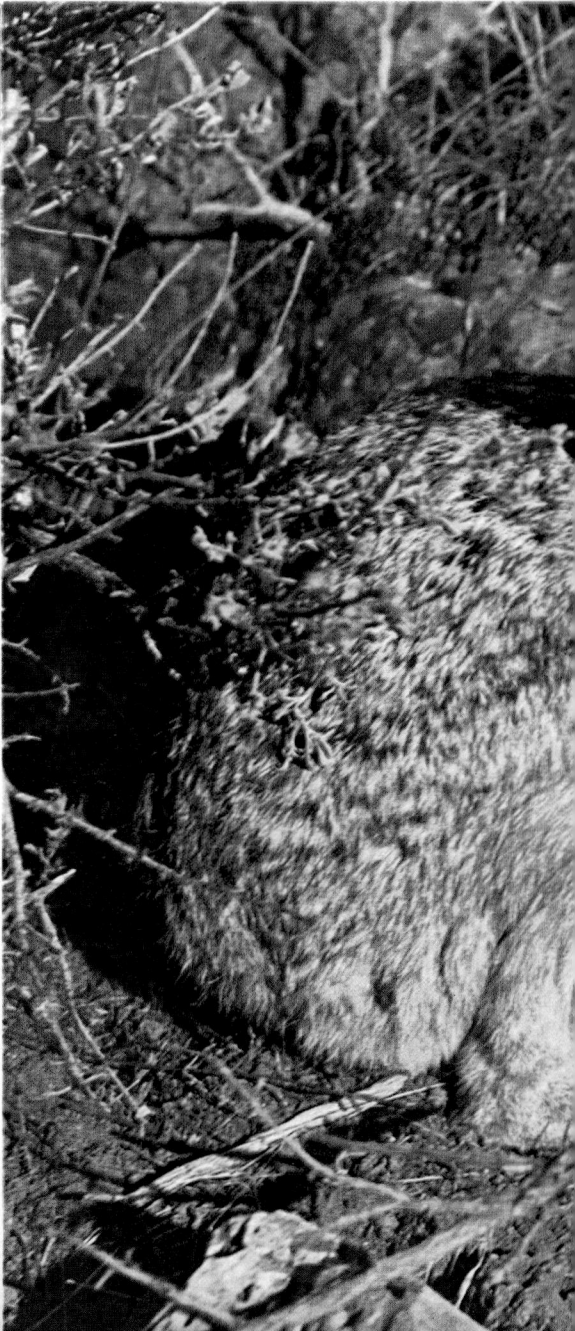

194

◀ **Jack Rabbit** *Lepus townsendi*
The jack rabbit is a hare rather than a rabbit, its closest relatives being the brown hare, varying hare and snowshoe hare. There are several species of jack rabbit, a native of the prairies of the western United States and Canada. Shown here is the white-tailed jack rabbit. Jack rabbits are well adapted to their grassland life, the two most obvious adaptations being well illustrated here. The long ears are used for detecting predators while they are still at a distance, while the long and powerful hind legs allow the animal to escape.

▲ **Pika**
The pika is a small relative of the rabbits and hares. There are 14 species, two found in North America and 12 in Asia. They live in a variety of habitats. One species, found on Mount Everest, lives at the highest altitude of any mammal. Others are found on all the plains, deserts, forests and rocky mountains. During the summer months they store food for the winter, cutting herbs, grasses and other vegetation and laying it out in the sun to dry. Later they store it in sheltered areas or burrows to provide a supply of food for the winter months.

▶▶ **Rabbit**
Oryctolagus cuniculus
The European rabbit was originally native to southwestern Europe and northwestern Africa, but its range has been extended considerably by man. Early Europeans realized its value as a domestic animal, and it was introduced into many parts of the world, where escaped rabbits formed large breeding populations in western Asia and North and South America. The most spectacular rabbit population explosion was seen in Australia, where the virtual absence of natural enemies allowed the rabbit to become a serious economic pest.

Wildlife of the sea

There is no part of the earth's surface that is as rich or as profitable as the sea. It teems with animal life, from its uppermost layers to its deepest parts. Because the sea is composed of a single element, water, it is often thought of as somehow uniform. It is no more uniform than the land, whose element is air. Land animals extract oxygen from the air, but this is in no way connected with the variety of ecological habitats found there. So it is with the sea. Marine animals extract their oxygen from the water, the medium in which they live. But each species of marine animal is specialized to live in its particular watery habitat.

What variations are found in the environment of the sea? Perhaps the amount of light is the first thing that springs to mind. A lot of light is present in surface waters, but the depth to which it can penetrate is only slightly variable, and at a depth of 20 fathoms there is not enough light to support plant life. So some animals are adapted to living in the light, others adapted for a sort of 'twilight' zone, and others spend most of their lives in darkness. The sea also has a wide range of temperature variations, not only the more obvious examples, like warm tropical waters as opposed to cold polar regions, but temperature variations according to depth – warm on top, cold on the bottom, with distinct layers of different temperatures in between. There are the familiar warm and cold currents – the rivers of the sea which have a tremendous effect on local areas. Then there are the variations in the ocean floor – muddy, sandy, rocky, coral-covered and plant-covered. There are undersea mountain ranges and gorges. There are the biologically rich estuaries, where

fresh water meets the sea, bringing with it soil, minerals, plant and animal life, and nowadays, pollution.

This will give some idea of the tremendous variations found in the seas of the world. It is therefore not so surprising that the animal life should be so rich and varied. A typical trawl from the North Sea, for example, would yield cod, haddock, whiting, dogfish,

Butterflyfish and Damselfish
This photograph shows two butterflyfish and two damselfish below them. All of them are relatively small bony fish which inhabit tropical waters in and around coral reefs. They are typical of the brightly-coloured inhabitants of this type of environment, and show the bright colours and bizarre patterns exhibited by so many marine animals.

◄ Ringed Seal

The ringed seal is one of the most common seals, and one of the most populous of the marine mammals. It is an inhabitant of the Arctic seas. Its numbers are estimated to be about six million and it has close relatives in the Caspian Sea and Lake Baikal. They prefer fast ice to loose floes, and herds are constantly on the move, seeking cooler water. They eat some fish, but for the most part live on plankton and small invertebrates. They have many natural enemies – polar bears, walruses, Arctic foxes and killer whales – and the young are especially at risk. Ringed seals are also important to the economy of the Eskimo, both for fur and food. The white coats of the pups are especially prized.

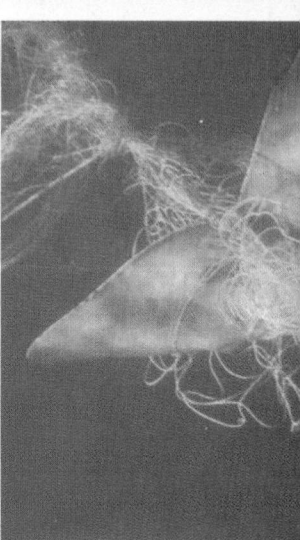

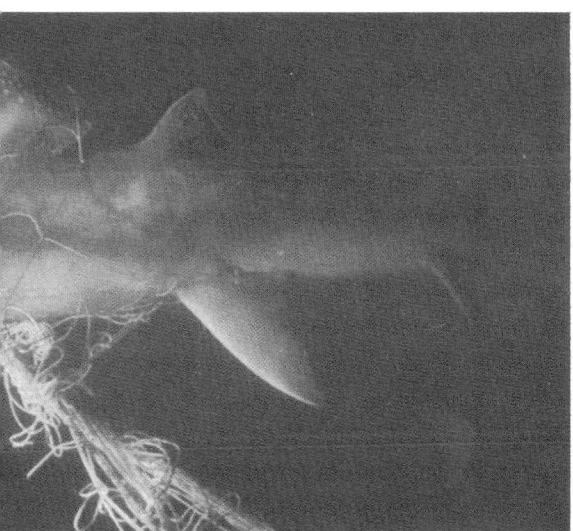

◀ **Butterfish** *Pholis gunnellus*
The butterfish takes its name from the fact that it is a very slimy and slippery fish, and therefore difficult to hold. It is a teleost and a native of the coastal waters between the tidemarks of the North Atlantic. Although it is rather eel-like in shape, they are not related.

◀ **Guitarfish**
The guitarfish is a not untypical member of the cartilaginous or elasmobranch fish. This photograph shows one caught in an anti-shark net. The guitarfish has a shape somewhere between that of a shark and that of a ray, but in habits it is much more like the latter, in that it is a bottom dweller, harmless and of little economic importance.

rays, catfish, whelks, cockles, pollack, skate, mackerel, turbot, plaice, gurnard and so on among the fish, and sponges, starfish of fantastic variety, scallops, sea urchins, sea anemones, crabs, lobsters and many other invertebrate animals. And of course a list like this of familiar creatures of the fisheries, does not include animals like sharks, turtles, seals, whales, penguins and many more marine animals.

Animal life in the sea follows the same general pattern as animal life on land. Marine animals are either herbivorous (plant eaters) or carnivorous (meat eaters); a few, the omnivores, eat both. Logically herbivores must outnumber the omnivores. But, in most places light does not penetrate to the ocean floor in sufficient amounts to support life, so what do herbivores feed on? The answer is to be found in plankton. The plants of the open sea, not of the sea bed, are microscopically small. They are suspended in countless billions upon billions in the upper layers – the top ten fathoms or so – where there is enough light for them. Most of them are composed of only a single cell, and they are able to float in the upper layers; a few have twin organelles which either support them or help them to move. These tiny plants provide a food source for vast numbers of small animals of many different kinds – worms, crustaceans, molluscs, the larvae of many kinds of larger sea animals, including fish – all of them no larger than the smallest terrestrial insects. This incredible mass of living organisms is called collectively, the plankton, derived from a Greek word which means 'that which is made to wander'.

◀ **Humpback Whale**
The humpback whale is one of the larger whales. The whales themselves are the largest living mammals. The humpback gets its name from its posture as it arches its back as it dives.

▲ **Sea Urchin**
The sea urchins, one of the most common echinoderms, are found in all marine waters around the world. There are more than 800 species of sea urchin, many of them large and coloured.

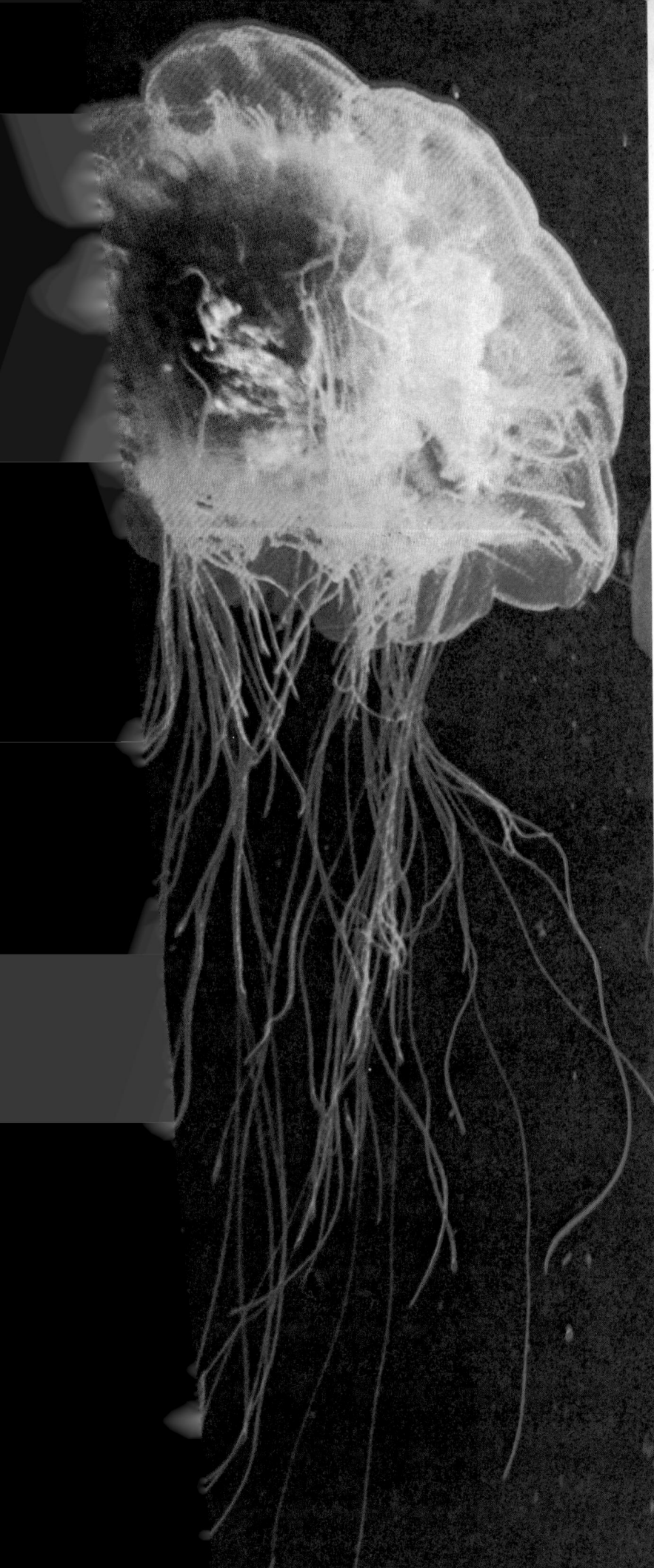

Coelenterates

The coelenterates are a large group of fairly primitive aquatic animals, most of which are found in the sea, although there are a few found in fresh water. Common examples of coelenterates are the jellyfish, the sea anemones and the corals. The coelenterates were an important step in the evolution of animal life. They were the first group to develop tissues – aggregations of cells which perform different functions and co-operate for the general well-being of the animal. This is a significant advance over the primitive single-celled protozoa, in which all physiological functions are carried on by the one cell, and over the first multi-cellular animals, the sponges, which are for the most part aggregations of single cells.

In the coelenterates, the body of the animal is composed of two layers of cells, an outer protective epidermis and an inner lining, the endodermis. Between these two layers of cells is a non-cellular middle layer which is secreted by the epidermis and the endodermis. Certain cells in the bodies of coelenterates have become specialized to perform specific functions. Some cells have become gonads and produce the sex cells. Others are primarily for digestion and secrete enzymes into the central cavity of the body of the animal. Cells of the corals have developed the ability to secrete limestone, building up huge underwater reefs as the colony grows and expands. Certain cells of the epidermis have become specialized for defence and for the capture of prey. These are called nematocysts and are cells which contain a tiny dart at the end of a coiled thread. When the cell is stimulated, usually by touch, the dart flies out of the cell and penetrates the object which stimulated it. Poison from the cell body streams down the thread and into the object, which is usually prey or a predator, stinging and anaesthetizing it. There are hundreds of thousands of these nematocysts on a single coelenterate. Most of them are concentrated on the tentacles which are organs specialized for defence and for the capture of food, and are located in a circle around the animal's mouth.

There are three main groups of coelenterates, which are identified by their structure and their habits. The first two of these are the hydrozoa and the scyphozoa.

◄ Jellyfish

The jellyfish is the common name given to the free-swimming members of the coelenterates. In the Scyphozoan coelenterates, the jellyfish, or medusa stage, is the dominant one. Some jellyfish may be very large. This is a photograph of a young so-called sea blubber. This particular animal can grow to the size of a man and the tentacles may be up to 33 times as long again, but it is more usually about half that size when found outside Arctic waters.

When a jellyfish is washed up on the shore, its body does not remain long under the rays of the sun. This is because it is practically all water. In fact, the body contains less than five percent organic matter and about 99 per cent of the animal consists of jelly.

Most coelenterates are small animals, but in many species individual animals are joined together to form larger colonies. The individual members of a colony are called zooids, and they are of two main types. Hydroids or polyps are specialized for feeding, while the medusae are reproductive. In the hydrozoan coelenterates, the mainly sessile polyps, that is, those attached directly by the base, are the dominant stage of the life cycle. They grow and reproduce asexually, but also produce medusae, free-swimming animals which have either male or female gonads for sexual reproduction. In the scyphozoan coelenterates, the medusae are the dominant stage of the life cycle, the polyps being short-lived and inconspicuous.

The third group of coelenterates are the anthozoans. They have no medusae and the polyps become sexually mature. They can be either large individual animals, such as sea anemones, or colonial aggregations of small ones, such as corals. However, none of these descriptions is completely accurate for every species of each group, as certain stages in the life history may be suppressed or absent altogether, as in the Portuguese man-o'-war, a hydrozoan and one of the most complex of the coelenterates.

Coelenterates are for the most part sedentary. Those that are not have only limited swimming powers and drift with the currents. In some coelenterates, muscle cells, such as the tentacles, have been developed, or limited movements of the whole animal are possible. They do not move about in their search for food but trap what comes near them. Any movement they do achieve is brought about by regular pulsations of the bell, the umbrella-shaped part of the animal. This is controlled by a very primitive nervous system and a series of organs around the bell which are sensitive to the pull of gravity, to light and to any chemical substances in the surrounding water.

◄ **Portuguese Man-o'-War**
The Portuguese man-o'-war is one of the most beautiful jellyfish. There are two species of the genus Physalia, one found in the Atlantic, the other in the Pacific. This photograph shows the typical structure. The animal is composed of a colony of four different kinds of polyp. A large bladder-like gas-filled float is the first, usually brightly coloured, under which hang the other three, each type specialized to carry out the particular functions of prey-catching, feeding and reproduction.

►► **Anemone**
The so-called sea anemones were named by investigators who were fooled by their appearance and took them for plants. They have long since been correctly identified as coelenterates, but the plant-like name has persisted. This photograph gives a good idea of the structure of one type of anemone. There is a central stock or body of the animal, which is attached to a suitable solid support by its base. At the other end is the mouth, surrounded by a ring of tentacles which wave gently to and fro in the currents. When suitable prey – a shrimp or small fish, for example – touches one of the tentacles, it is immediately stung and paralyzed by stinging cells in the tentacles. Other tentacles move into action, stinging and holding, slowly conveying the prey towards the mouth, where it is ingested.

◀ **Portuguese Man-o'-War** *Physalia physalis*
This is a photograph of a brightly-coloured specimen
of the Atlantic Portuguese man-o'-war. Its large
inflatable float causes the animal to be blown along the
surface of the water, so although it has no locomotory
ability of its own, it is still quite mobile. The bladder is
filled with a mixture of gases of much the same
composition as air, although lighter, which are secreted
by a gas gland. Muscles in the coat of the bladder
regulate the pressure inside. The bladder is deflated in
storms and the jellyfish rides out the weather beneath the
surface. The float can be re-inflated in a matter of a few
minutes. In rough seas the tentacles are spread out in a
circle around the bladder to help maintain the animal's
balance. The float is always twisting and turning,
dipping below the surface with the action of the waves,
keeping its outer surface moist.

The Portuguese man-o'-war is well known for its
stinging tentacles. The stinging darts or nematocysts,
tiny organelles which are kept in capsules, are located
on the prominent circular rings on these tentacles.
Swimmers who have come in contact with these
tentacles will testify that the darts are capable of
producing an inflammation of the skin similar to that
of a severe burn. It has been estimated that the poison
in a man-o'-war's darts is 75 percent of the strength of
cobra venom, so it is no wonder that they produce so
violent a reaction.

▶ **Sea Fir** *Obelia geniculata*
This colonial hydroid is a common seashore animal, but
is often mistaken for a seaweed. This photograph of one
of the more than 2000 species of this animal, magnifies
a portion of the colony about 60 times, and shows its
structure well. Members of the colony are connected
together by a branching stem, through the centre of
which is a system of intercommunication. The
individual polyps of the colony are the cup-shaped
structures which are supported by stems arising from
the main branch at one end, and provided with a fringe
of tentacles around the mouth at the other end. These
tentacles are mobile and provided with stinging darts, as
in other coelenterates, and are used for catching prey.
Because of the system of intercommunication between
individual members of the colony, it is necessary for only
some polyps to catch prey in order to feed the whole
colony.

Several features are seen in this photograph. The
circular structure with tentacles just below the polyp
at the top-right is the medusa phase of the animal. This
is a free-swimming form, either male or female, which
has been produced by a special sexual generation-
forming apparatus like the one seen immediately to its
left attached to the stalk. The medusa swims away from
the colony and sheds its sex cells, ova or sperm, into the
sea, where they come into contact with their opposite
numbers from other medusae. The fertilized eggs sink to
the bottom and begin to grow into a new colony.

Also seen here are tiny, stalked single-celled
organisms, which have chosen the sea fir as a suitable
base to which to attach themselves.

▲ Millepore *Millepora alcicornis*
The millepore is sometimes called the millepore coral because of its superficial resemblance to the true corals and because it is often found on coral reefs. The millepore is made up of colonies of tiny hydroid-like polyps supported by a hard chalky skeleton. Millepores usually form upright columns which branch coarsely. To some extent, the shape of the colony depends on the base on which it lives. It can be plate-like when growing on rocks or similar smooth surfaces, or upright and branching if there is the necessity to grow upwards.

The surface of the millepore is covered with a thin layer of living tissue from which arise thousands of tiny polyps. The polyps can be retracted, if necessary, into tiny holes, and it is the profusion of these openings in the skeleton of the animal that gives it its name of millepore or 'thousand pores'. There are two quite distinct types of polyps, one a short, stout, feeding variety with a mouth and four or more knobs armed with stinging cells, while the other is long and slender, without a mouth but bearing long tentacles with knobs armed with stinging cells for capturing prey. Large stinging cells are also found on the 'body' of the colony. Some species of worms, crustaceans and algae live on millepore colonies.

▶ Anemone
This photograph illustrates one of the many mysteries about the coelenterates. It has been noted that fish coming into contact with the anemone's tentacles cause activation of the stinging cells and the movement of the tentacles inwards towards the mouth to secure the prey and ingest it. It has been determined that both touch and chemical stimuli can cause this activation. Yet here are three clown fish swimming quite freely among the tentacles of an anemone. It appears almost as though the anemone has accustomed itself to the presence of the fish, although the mechanism by which this could happen is hard to imagine. It is possible that the clown fish has developed some sort of protective device that, rather than making it immune to the stings of the anemone, prevents them from discharging in the first place.

Sea anemones are surprisingly mobile. They can grasp the substratum with their tentacles and release their grip at the base, thus being able to move along the ocean floor by a kind of somersault movement. They can also just let go and float away until they settle in a new place. There are even several species of anemones that burrow in the sand or mud. They can survive for a considerable time without food.

Soft Coral

Corals are colonial organizations of polyps, each of which is similar in shape and structure to an anemone, except that it is much smaller. They are supported by an extremely hard, white, calcareous skeleton which, in a dead coral, is white, but which in the living animal takes on the brilliant coloration of the particular species of coral which has deposited it. There are several different types of corals, some of which are called 'tree' corals, but these different groups of coelenterates are separated by relatively minute differences. Basically, there are two types of corals, the reef corals, which form enormous underwater reefs like the 2000 kilometre-long Great Barrier Reef of Australia, and the so-called soft corals, which are illustrated here.

The term soft coral is misleading because only a few corals have any degree of flexibility. The only reason for using the term at all is that it has been used frequently in text books, although even this is not particularly helpful, as it appears that different authors are referring to quite different things when they write about soft corals.

The term soft coral was first used by the Greeks and the Romans. They believed that corals were soft like sponges, when living in the sea, but hardened when they were brought into the air. There are no coral reefs in the Mediterranean, and presumably those which were brought up were in fishermen's nets, so perhaps the confusion is understandable. The first coral to be called a soft coral was Corallium rubrum (illustrated below left), the red or precious coral. Its bright red branching colonies are covered with white polyps with long tentacles, and with smaller polyps without tentacles which lie partly hidden between the tentacular forms. This photograph gives some impression of the wide range of colour, structure and complexity exhibited by these animals.

Most soft corals live in seas from depths of a few millimetres to more than 2000 fathoms although most species are found at 500 fathoms or less. They especially prefer a rocky sea bed for attachment, and are most numerous in the warmer seas, especially the warmest areas of the Indian and Pacific oceans. The size and structure of individual colonies is extremely variable. The colony may be small and low-lying or it may be up to two metres high. In areas in which there is a more or less constant current, the colony will probably be fan-like, its broad side arranged at right angles to the current so that the maximum surface of the colony is exposed to the passage of water containing potential prey. In deeper waters, where there are few currents, the colony may

branch in all directions, or form large irregular masses.

The tentacles of each polyp trap any prey that swims into contact with them. The prey is quickly paralyzed by the stinging cells and passed to the mouth. As each polyp is connected to its neighbours, the entire colony can feed on the food caught by a few members, so it is not necessary for each polyp to do its own feeding.

Reproduction, as in many other coelenterates, is both sexual and asexual. Sexual reproduction, in which an egg is fertilized by a sperm, gives rise to a free-swimming larva which moves away from the colony, settles on a suitable substratum and grows into a polyp. Further polyps are produced by asexual reproduction or budding. A bud appears on the side of the polyp which slowly develops into a polyp of the same size as the parent. This process is repeated, and so the colony grows. If for any reason a part of the colony should be broken off, or a single polyp become detached, it will continue to grow and will form a new colony.

Some corals have in the past been regarded as semi-precious. These are particularly the ones in which the hard calcareous skeleton of the colony retains its colour after the death of the polyps, rather than taking its colour from the living polyps. The best known of the semi-precious corals is the red coral of the Mediterranean. It is composed of branching, tree-like colonies. The branches are made of hard red spicules which are firmly cemented together. At one time there was a flourishing industry fishing for the red coral in the Mediterranean, the Red Sea and the Persian Gulf. The red coral lives at depths of from two to one hundred fathoms, so was not difficult to harvest. It was used in commerce and for the making of jewellery.

The so-called tree corals differ from those illustrated here in several ways. They are supported by a hard skeleton which, in life, is covered by a continuous layer of tissue giving the coral its colour. The true or stony corals may be either solitary or colonial. In both cases, the polyps lay down beneath them the hard skeleton which is typical of the corals, in a pattern which is distinctive for each species. The true corals or reef corals are mostly night feeders. During the day the polyps are withdrawn and the surface of the colony is smooth, but at night, as the plankton rise towards the surface to feed, the polyps take in water, swell and stand out on the surface, forming an intricate meshwork of tentacles which trap prey. As in the soft corals, prey is paralyzed by stinging darts, and conveyed to the mouth by the tentacles and also by the cilia which may cover the surface of each polyp.

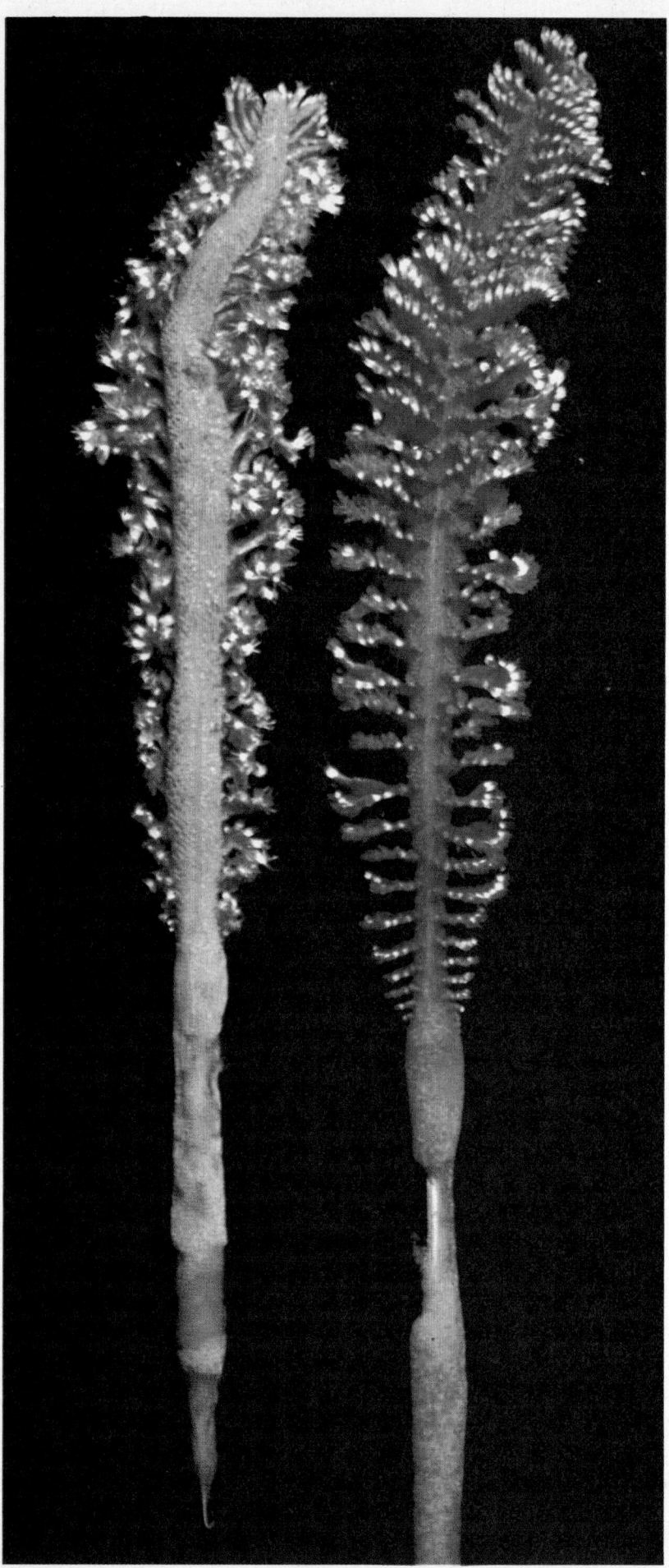

◄ Sea Pen

The sea pens derive their name from the fact that they reminded their discoverers of the old-fashioned quill pens. In fact the sea pens come in a variety of shapes and sizes, some having the common name sea pansies. Basically, though, regardless of shape and size, they all have the same form: a central stem from the top of which protrude polyps or branches bearing polyps. The sea pens begin life as a single polyp which develops from a free-swimming larva. This forms the central stem, and from this bud the secondary polyps. Usually the mouth and tentacles of the primary polyp degenerate, leaving only a stem-like structure which supports the secondary polyps and gets its nourishment from them. A central rod supports the stem and the polyps lay down a complex skeleton of interconnecting spicules which help the colony to remain upright.

The sea pens are close relatives of the anemones and sea firs, and therefore the structure of their polyps is very similar. One peculiar characteristic of the sea pens is their ability to give off light when touched. The light is caused by the release of mucus filled with luminous granules, but the reason for this display, which can be spectacular in a forest of colonies, has not been explained.

► Portuguese Man-o'-War

This photograph gives some indication of the size of prey that this animal can cope with, and is also a particularly good illustration of the crest on the bladder, which acts as a sort of sail when the man-o'-war drifts with the wind and the currents.

Like other coelenterates, the colony begins with a single motile larva. This larva develops into a small bladder with the gas gland at one end. At the other end, the mouth appears, forming the first feeding polyp. Between the two ends the budding zone develops. The first growth from this zone is a stinging tentacle, and this is quickly followed by more feeding polyps and stinging tentacles. Later, the reproductive polyps, some male, the others female, develop. In the meantime, the bladder has continued to grow, and has developed its crest. Growth continues from the budding zone, giving rise to stinging, feeding and reproductive polyps until the animal is ready to breed. Eggs and sperm are shed into the sea, and the colony probably dies at this stage, still only a few months old.

The stinging tentacles may be very long. The darts which they release when prey comes in contact with them penetrate the skin of the prey, fixing it firmly and at the same time providing a tube down which poison can pass to kill the prey. The stinging tentacles are contractile, and once they have captured something, they draw it towards the mouth. As the prey nears the main mass of the animal, the flask-shaped feeding polyps begin to move about. When they touch the prey, they fasten on to it, sucker-like, the many mouths of the feeding polyps completely enclosing large animals and pouring digestive enzymes over it. Small prey can be ingested by a single polyp.

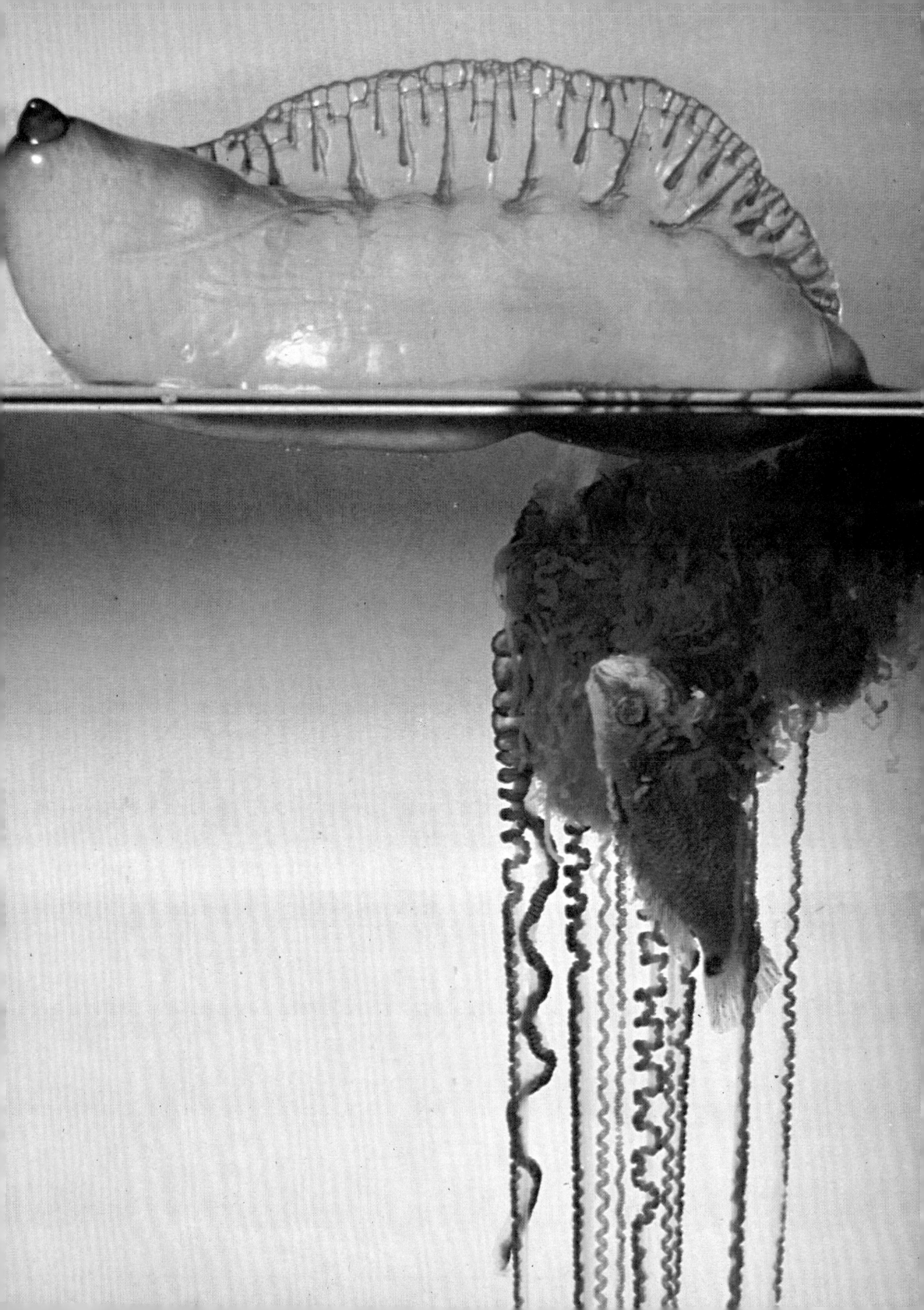

Crustaceans

The crustaceans are some of the aquatic members of the arthropods, the largest group of animals both in numbers of species and of individuals. The arthropods are segmented animals whose bodies are divided into three distinct regions – head, thorax and abdomen. They have a hard external covering, the exoskeleton, which performs many of the same functions as the bony internal skeleton of the vertebrates.

Crustacean arthropods are distinguished from other arthropods by the presence of two pairs of frontal antennae and by the nauplius larva. Crustaceans lay eggs which hatch to produce the nauplius larva, a simple unspecialized sort of crustacean common denominator. Only as it grows and develops is it possible to tell what sort of crustacean it will eventually become.

Crustaceans are much more diverse than is usually assumed. All are aquatic and most are marine, but there are a few fresh-water types such as the cray-fish and the water flea. Marine crustacea include not only the common lobsters, crabs and shrimps, but less obvious members such as fish lice, beach fleas and barnacles, the last often confused with molluscs because of their shell-like exoskeletons. There are even crustacea which have become specialized as internal parasites of other crustacea.

Like other arthropods, crustaceans have to moult from time to time while they · are developing and increasing in size. The hard exoskeleton is made of a substance called chitin, which is secreted by glands in the animal's skin. It is not living tissue and cannot stretch to accommodate the increase in size of the animal, so the exoskeleton must be shed. A special secretion called moulting fluid causes the exoskeleton to separate from the epidermis, the outer layer of skin, beneath it. Much of the inner portion of the exoskeleton is digested by special enzymes and is used in the production of the new exoskeleton, which is laid down underneath the old one. At this stage the new exoskeleton is soft and pliable. When all is ready for the moult, the crustacean takes in an enormous amount of water. This causes it to swell and break the now weakened exoskeleton along the middle of the back. The animal climbs out of its old suit of armour, continuing to take up water and swell to enlarge the new exoskeleton and allow for future growth. When it has reached its maximum size, enzymes are secreted by the epidermis to harden the chitin. In many crustacea, limestone from special glands is also added to the chitin at this stage to provide additional hardening. During the moult the crustacean is at its most vulnerable, but once the exoskeleton has hardened it is again well protected from predatory attack.

Squat Lobster

There are four species of squat lobsters, and in spite of their name, they are more closely related to the hermit crabs than to the lobsters. All four species are found in Europe. They live in the tidal zone or in the shallow offshore water, and are very small.

◄ **Prawn**

The prawns are small shrimp-like crustaceans which live in shallow inshore waters. The name was first applied to the common prawn found in European waters, but it has since come to apply to any small, long-bodied crustacean of marine waters. They spend most of their time walking slowly along the sandy inshore bottoms, for like so many of their crustacean relatives, they are scavengers, feeding on any dead animals that they can find, and occasionally taking a bit of seaweed. Some prawns are active hunters, however, hiding in burrows until suitable prey swims past and then either catching it with their large claws or delivering it a blow to stun it, then picking it up. The prawn's most obvious enemy is man, who fishes for it commercially, but most bottom-feeding animals include prawns in their diet. Its principal natural enemy is the cuttlefish. These two photographs on the left are of a common prawn (above) and a so-called piebald prawn (below), one of the more brightly coloured crustaceans common in tropical waters.

▶▲ **Shore Crab** Carcinus maenas

This is the common crab of the western coasts of Europe and the eastern coasts of North America, where it is known as the green crab. Two other shore crabs live along the Pacific coasts of North America, but they are quite different species. The European shore crab is a tough, defensive animal noted for its pugnacious ways. They live in rock pools in the tidal zone and in the sea to a depth of about three fathoms. It is one of the few marine crustaceans commonly found in estuaries and can exist, apparently quite happily, in water which has only 15 percent of the salinity of sea water. It eats a wide range of food, for it is a predator as well as a scavenger, and will eat any kind of animal food that comes its way. The shore crab is particularly notable for having one of the animal kingdom's most complex biological clocks. Laboratory study under constant environmental conditions has shown that shore crabs have peaks of activity which coincide with the times of high tide and the hours of darkness. This means that the shore crab is most active during the periods when it is least vulnerable.

▶▼ **Fiddler Crab**

The fiddler crab is the common name applied to a group of similar-looking species which are common on tropical sandy or muddy beaches. Only the male has the enormous oversized claw. On the female, both front claws are of normal size. The large claw of the male is so big that it is useless for all practical purposes, and it is therefore assumed to be important for some sort of ritual display. Several theories have been put forward, including use as an attractant for females, especially as it is brightly coloured, and use as a territorial marker. The latter is probably the more likely explanation, as the claw is constantly flashed at other males during low tide, warning them to keep their distance. If a male ventures too close to another, battle ensues, and the large claw is used to flip the opponent over on to his back.

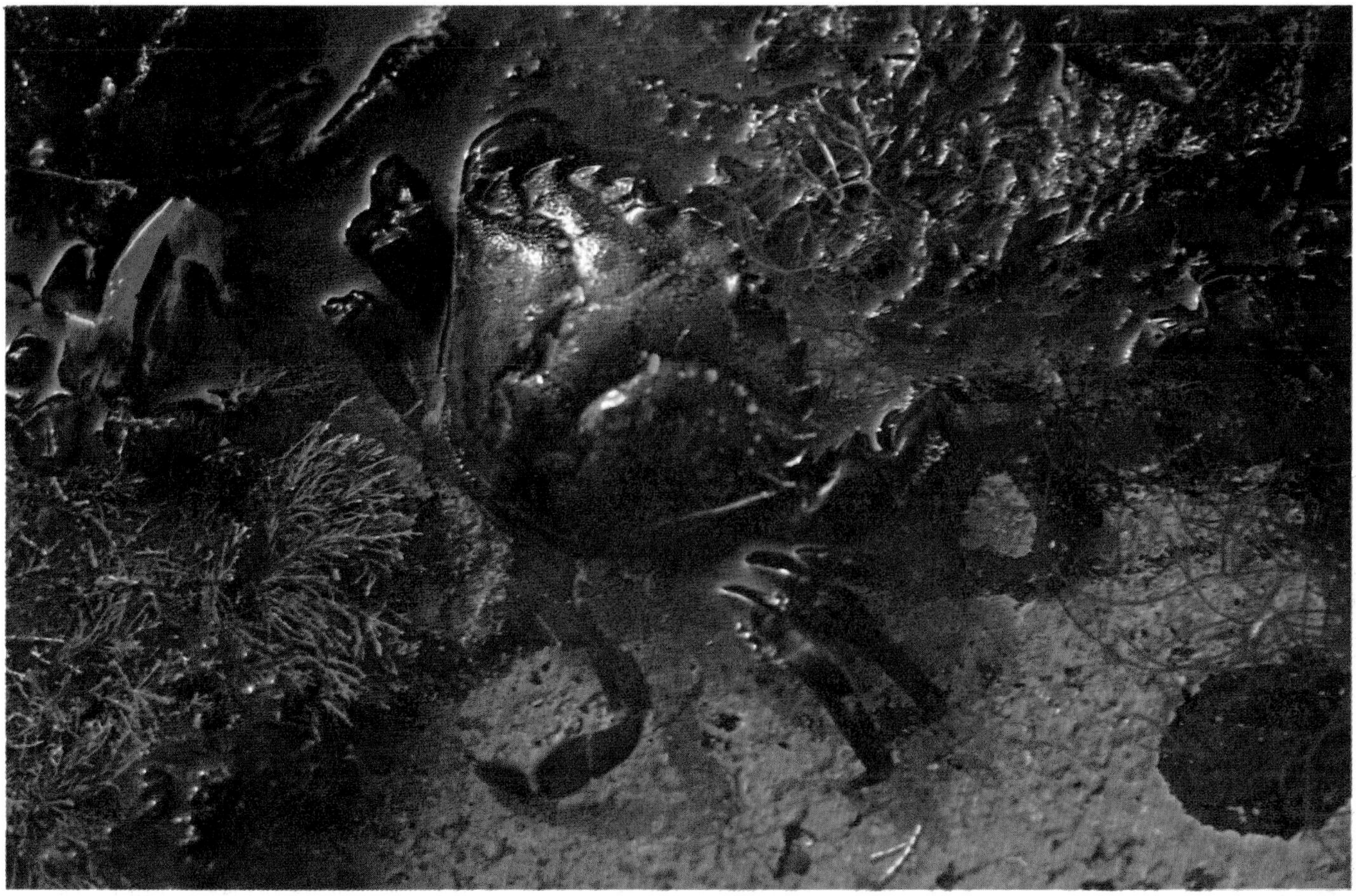

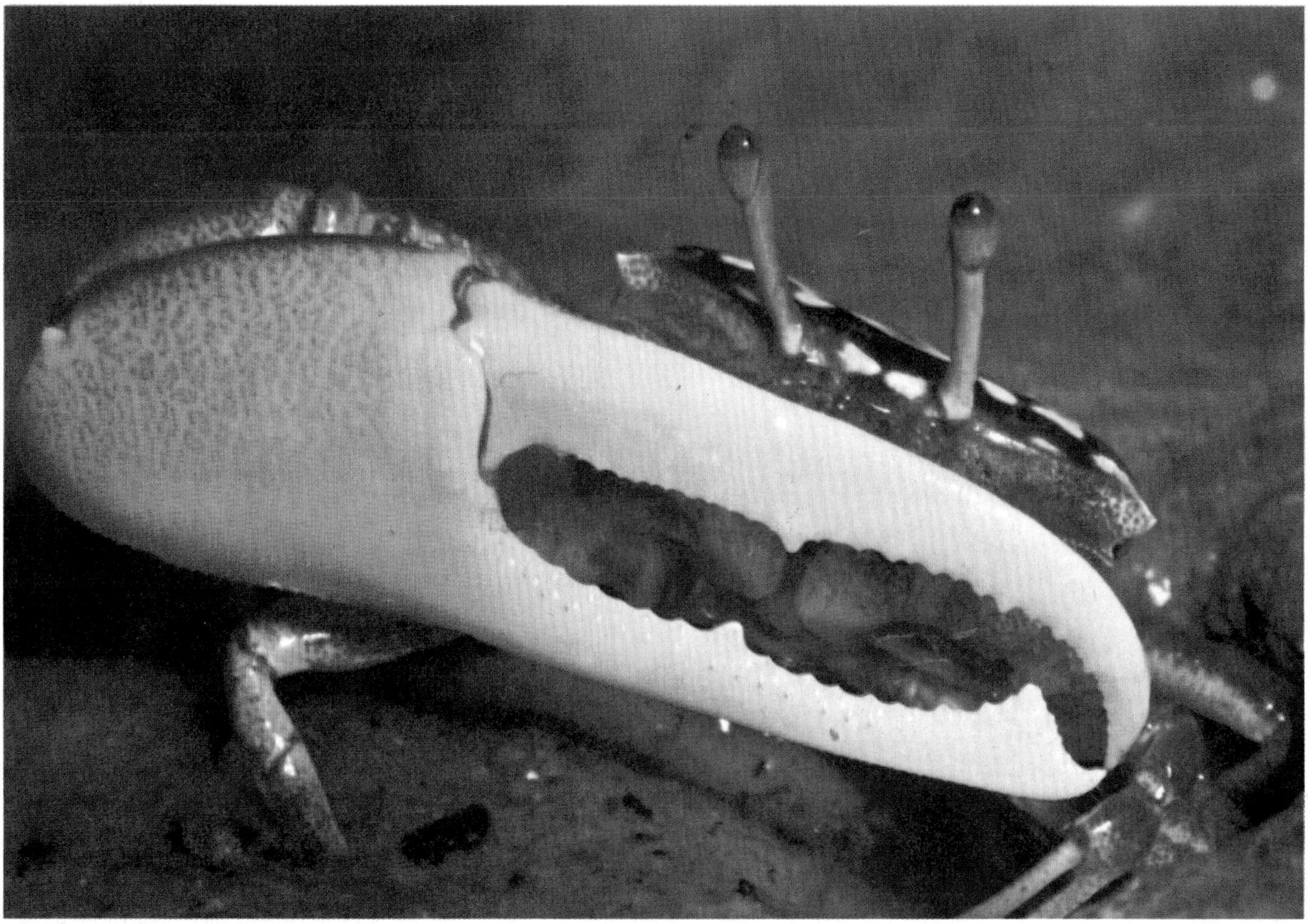

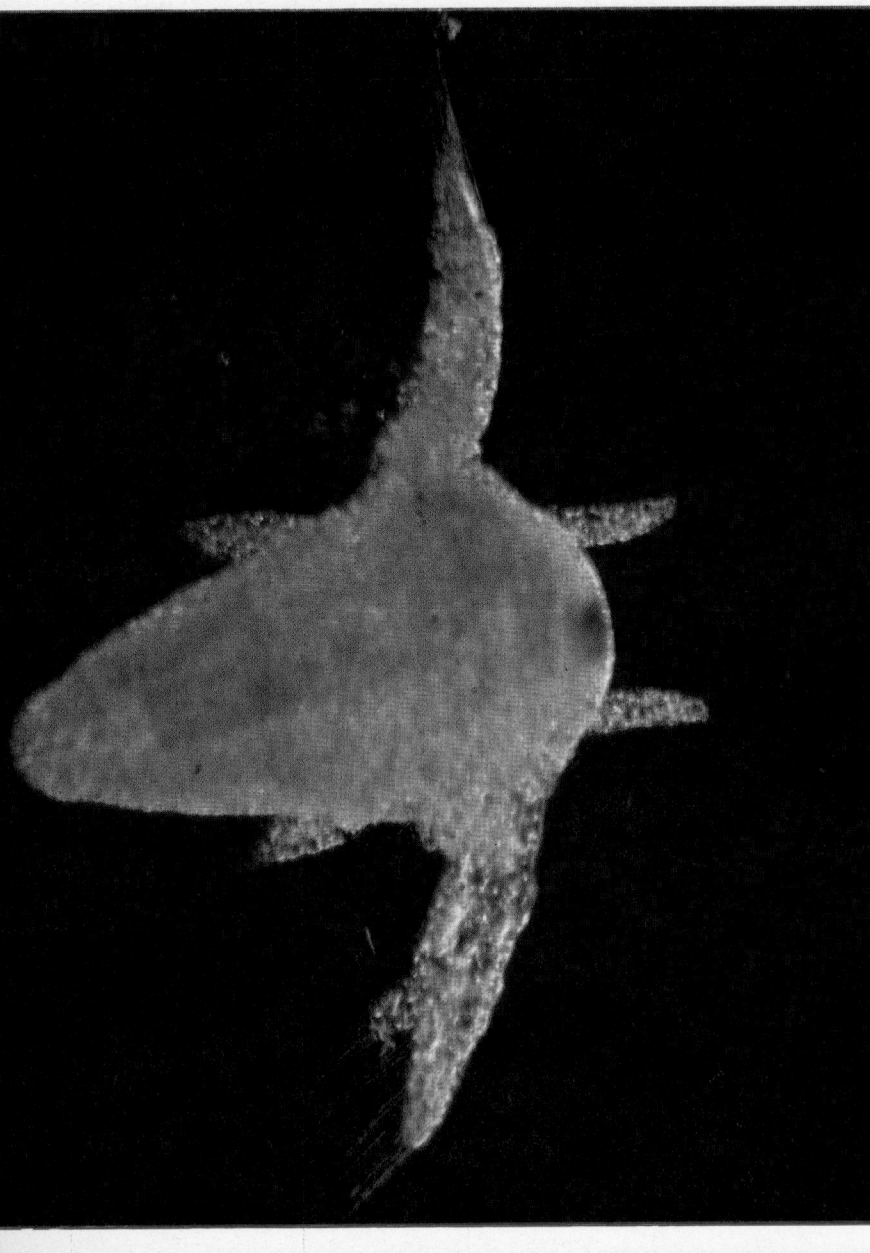

◄ Brine Shrimp *Artemia salina*
The brine shrimp is a small primitive relative of the crabs and lobsters. It lives in salt pools and other very salty waters in many parts of the world, on all continents. It is most common in water whose salt concentration is greater than that of sea water, such as pools, salt lakes and marshes. It feeds on particulate matter suspended in the water, the swimming actions of its 11 pairs of limbs sweeping a current of water towards the mouth where the tiny particles are caught in strainers made of bristles, located along the inner edges of the limbs. Brine shrimps can withstand enormous variations in their environment, from water of less than 0.3 percent salt up to 30 percent. The eggs of the brine shrimp are even more remarkable, for they will withstand complete desiccation, boiling and freezing, hatching a few hours after being placed in salt water at room temperature. This ability of eggs to withstand extreme conditions is doubtless of great survival value, as many of the ponds in which brine shrimps live are liable to dry up, exposing the eggs to the air and the hot sun for long periods until water becomes available again.

Reproduction is also a curious process in brine shrimps. The females have a special brood pouch at the posterior end of the body, and the young larvae, called nauplius larvae (below left), remain in this pouch for up to several weeks if conditions are not favourable to their survival. Under normal conditions, the larvae undergo a series of moults and become mature in about three weeks. In some populations of brine shrimps, males appear to be entirely absent, and reproduction is parthenogenetic, that is, the eggs develop into true adults without being fertilized.

Curiously enough, brine shrimps are not found in the open sea. This is probably because they could not stand the pressure from predators and have instead moved into an environment which most of their would-be predators would find too inhospitable. To cope with the increased salt concentration, the brine shrimp has developed a special physiological mechanism by which salt water is absorbed by the gut but the major portion of the salt content is excreted by special cells located on the gills at the bases of the limbs.

► Daphnia
Although not strictly a creature of the sea, as it is primarily a fresh water animal, Daphnia, the water flea, is worth a brief consideration, as it is one of the smallest crustaceans, the largest of the many species found all over the world being only a few millimetres long. It is a distant relative of the crabs and shrimps, with a hard external skeleton and jointed body and limbs. There is a large central eye at the front of the body, the result of the two eyes joining together during embryological development. The female carries the eggs in a brood pouch under the carapace, and does not lay them unless conditions are favourable. Like the eggs of the brine shrimp, they can survive adverse conditions and hatch years later if the environment is suitable.

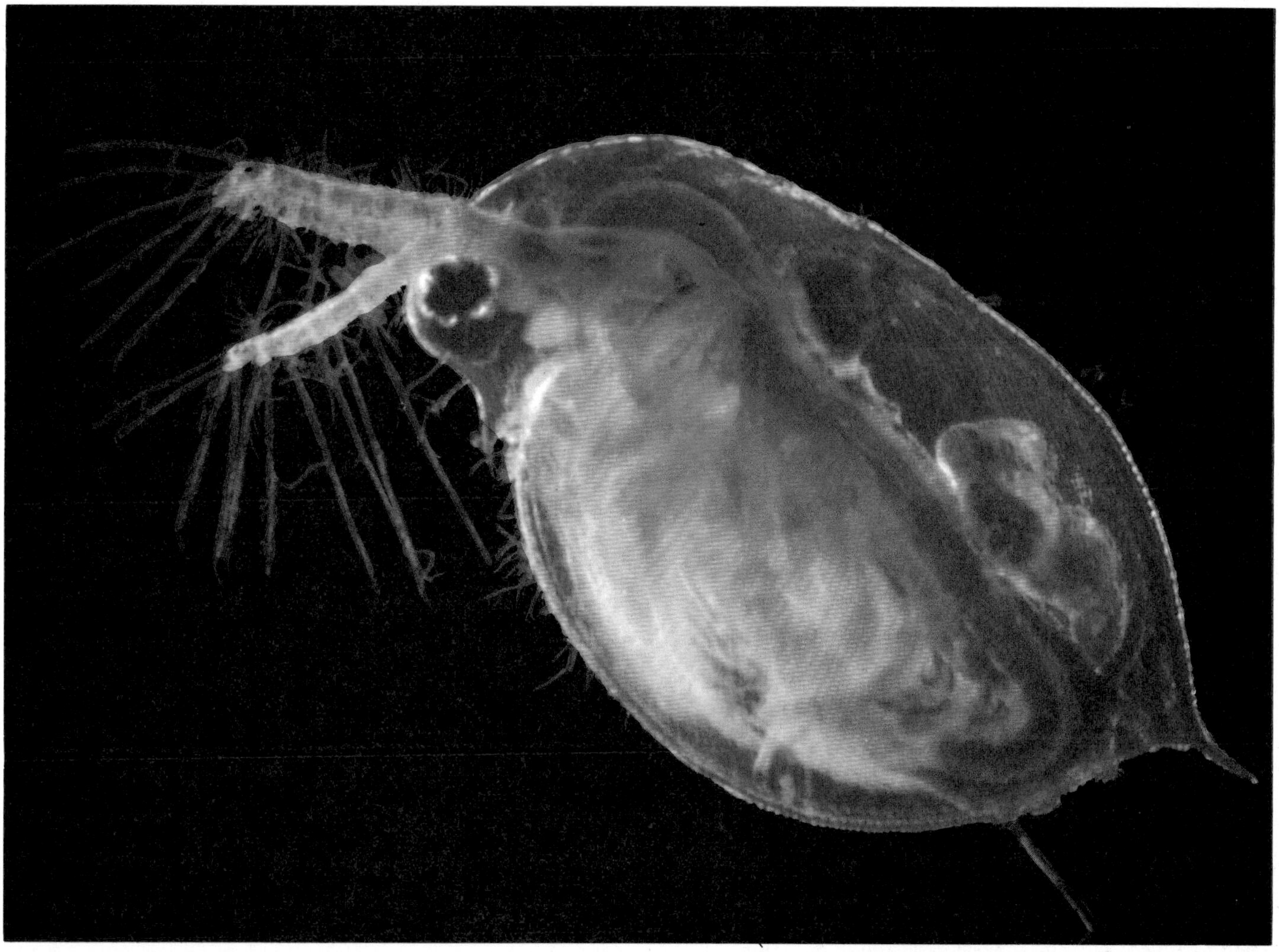

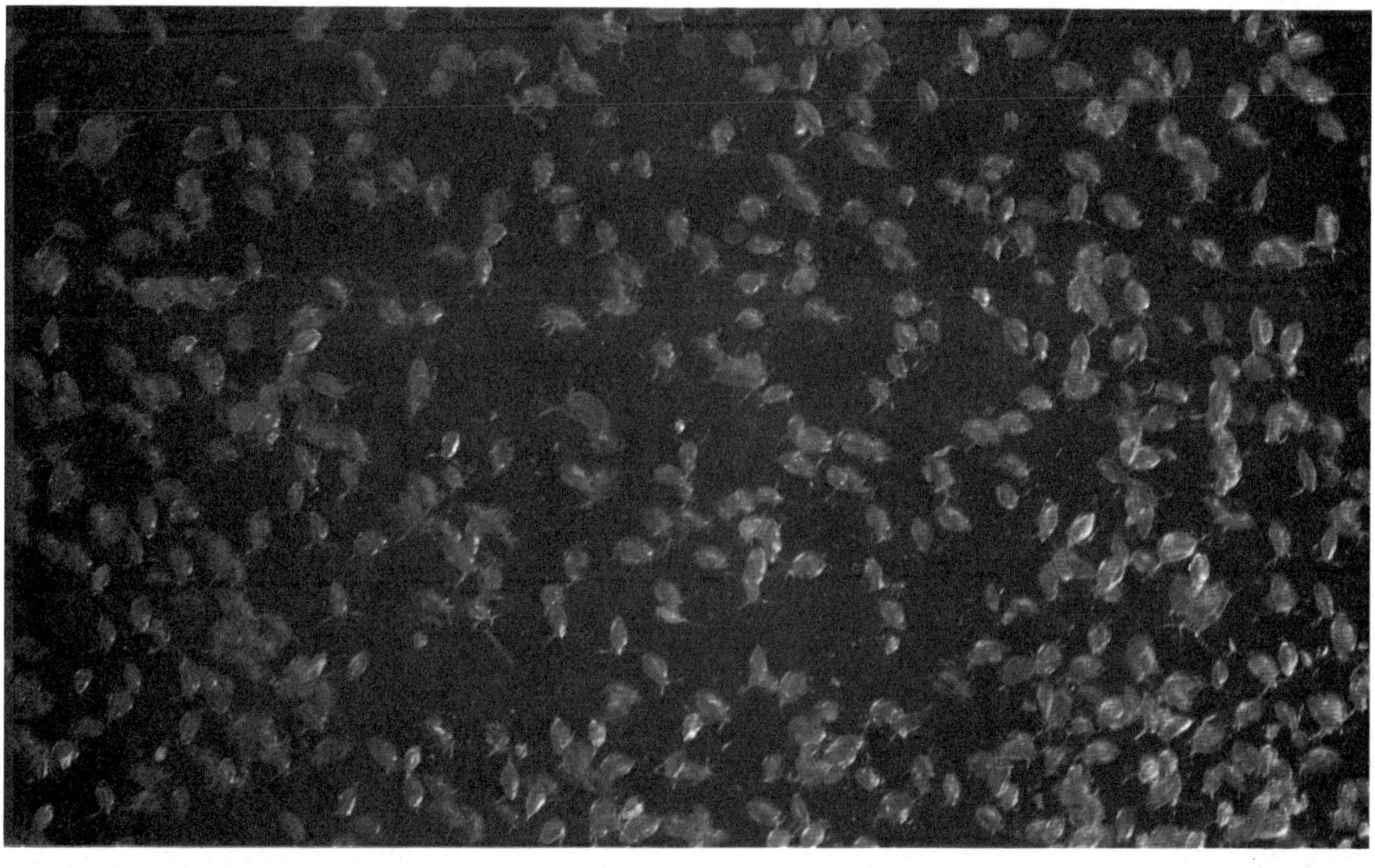

Hermit Crab

There are many different species of hermit crabs found in many parts of the world. Most are marine, but some tropical species are semi-terrestrial. They derive their name from their habit of choosing some alternative protective coat to carry about with them, rather than relying on the tough exoskeleton used by other crustaceans. The most common hermitages chosen by hermit crabs are the empty shells of snails. The shape of the crab's abdomen has become modified and its covering softened to allow it to fit easily into the shell. The front pairs of limbs have similarly become modified for this new way of life. One of the claws is much larger than the other, and is used to seal off the opening of the shell when the crab is threatened. The next two pairs of legs are used for walking. The next two pairs are reduced in size, and are used to grip the shell, while the last pair of abdominal

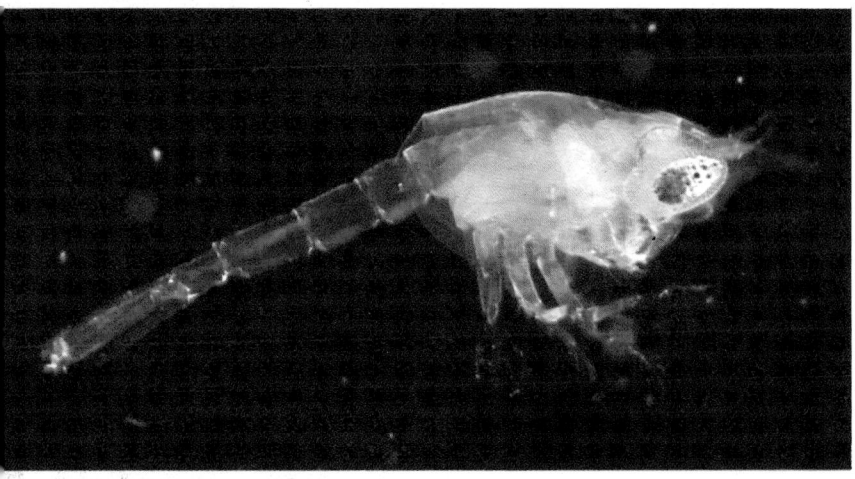

limbs, those which in the lobster form the fan of the tail, are specially shaped to grasp on the central column of the shell. These adaptations make it extremely difficult to remove the crab from its shell. The shell that is usually chosen first is that of a winkle, a topshell or a dog whelk. Normally only young hermit crabs are found on the shore. They are surprisingly nimble, in spite of their heavy load, and can move quickly across the sand. The shell also protects them from being pounded too heavily by the waves, and prevents their drying out. The picture below is a close-up of one of the more common hermit crabs, *Pagurus megistos*.

The four smaller photographs show different stages and aspects of the life cycle of the hermit crab. Most hermit crabs seem to be in breeding condition for most of the year. The female may carry as many as 10 to 15 thousand fertilized eggs, which are attached to abdominal appendages called swimmerets. The female will come part way out of her shell from time to time to aerate the eggs, fanning her swimmerets to and fro in the water. As soon as an egg hatches the young hermit crab moults and becomes what is known as a zoea larva (top), a tiny shrimp-like animal. This zoea larva lives as a free-swimming animal, undergoing a fairly quick succession of four moults. After the fourth moult, the young hermit crab, which now has the appearance of a miniature adult, seeks its first shell for a home. Once inside the shell, usually that of a species of snail at this stage, the crab is relatively safe from predators. Like all crustaceans, however, the crab must moult periodically as it increases in size. Increase in size means that the crab also outgrows its home and so it is necessary to find a new, larger one. Usually the crab finds a new shell before abandoning the old one, closely inspecting and appearing to measure the new shell as if to make sure it will be large enough. Once the decision is made, the crab leaves the first shell, a split appears along the back of the animal, and it wriggles out of its old skin. As quickly as possible thereafter, it transfers to the new shell, for at this stage the soft-bodied hermit crab is particularly vulnerable to predators. Such a naked crab is shown in the second photograph.

The third photograph on this page shows a hermit crab in the shell of a snail. Part of the shell has been cut away to show how the hermit crab grips the shell with its legs and wraps its soft abdomen around the central coil of the shell. The bottom picture is of the same hermit crab, the shell joined together artificially, apparently none the worse for wear.

Not all hermit crabs make their hermitage in a shell. One East Indian species seems to make use of bits of bamboo, coconut shells and even suitable bits of man-made refuse. One of the West Indian hermit crabs first inhabits a snail shell, but this quickly becomes overgrown by a species of bryozoan, a primitive marine organism, which dissolves the shell and remains to protect the crab and grow with it. Two other species, one West Indian, and the other from the Indian Ocean, make sticks of bamboo serve as their homes. These animals have straight abdomens with a modification to the posterior end which acts as a sort of plug. Still other species do not have mobile homes, but live in holes or crevices in coral or sponge. Some hermit crabs have been known to drive others from their shells in their search for a new home.

▼ Hermit Crab

Strangest of all are the hermit crabs of some tropical regions, which are partly terrestrial. The coconut crab makes a burrow at the base of a coconut tree and lines it with coconut husks. Some of these land-inhabiting hermit crabs may climb bushes in search of plant food, and are reported to rob eggs from nests and to attack young birds. Most hermit crabs are omnivorous, eating just about anything they can find, including a variety of plant food, live animals and carrion.

One of the strangest associations of hermit crabs is with particular species of sea anemones. One or more anemones may be perched on the shell of the crab, and some investigators have suggested that the crab may even go so far as to 'pick' the anemones and place them on its shell. In fact, at least one species of anemone is found only in this kind of association with a hermit crab. This partnership of crab and anemone is one of the better examples of communal living. It is not parasitism because both animals benefit from this association. The anemone picks up bits and pieces of food which the crab tears free from its prey, and is much more mobile because of its association with the crab. The hermit crab carries with it a built-in camouflage, and the stinging darts on the tentacles of the anemone provide an additional defence against would-be predators such as the octopus and squid. One species lives in water which is too deep for light to penetrate, although it has very large eyes. It is thought that the phosphorescent anemone it carries around with it provides enough light for it to find its way.

▶ Prawn

Some prawns are chameleon-like, able to take on the colour of their background. They are able to vary from green on seaweed on the sea floor, to red on red seaweed. By night they are blue, regardless of the colour of their background. This colour change is not a rapid one, as in some other colour-adaptable animals. The change may take up to a week and the prawn first attempts to find a background that is the same colour as its own, rather than adapt itself to its environment. It is only when it cannot do this that it changes colour. The colour change is brought about by hormones. Light entering the prawn's eyes also activates glands in the eyeballs. These are probably different kinds of glands which respond to different colours of light, and the hormones produced by these glands are released into the blood, stimulating the pigment cells in the prawn's skin. Thus red light would stimulate only 'red' glands, and the hormone released by these glands would cause the prawn to produce only red pigment in its skin.

As well as this remarkable ability, some deep sea prawns have light organs. These give out a sort of greenish-yellow light. They are scattered along the sides of the prawn, up to 150 of them on a single animal. The prawn itself may be quite big for some of these deep sea dwellers may reach a length of 30 cm or more. The light organs may flash on and off simultaneously, independently, or in series in a regular progression from head to tail. Man is one of the many predators of the prawn. They are also included in the diets of most bottom-feeding fish.

Molluscs

The molluscs, one of the groups of higher invertebrates, are characterized by having a soft body which is usually, but not always, covered by an external protective shell secreted by the animal. The ventral (stomach-side) of most molluscs is modified to form a large muscular foot for locomotion, and the tongue has a unique rasping device on it called the radula. There are five separate and distinct groups of molluscs. Two of them, the amphineurons, of which the chiton is an example, and the scaphopods, the so-called tooth shells, will not be discussed here. The other three are more familiar, some members having important commercial associations.

The gastropods are the only molluscs found on land as well as in the water. Snails, slugs, whelks, abalones and limpets are members of this group. There are both herbivorous and carnivorous gastropods, with appropriate modifications to the radula according to diet. Most of them move about slowly on their broad and muscular foot, but a few use it as a lever for jumping, while others use it as a fin for swimming.

Torsion, the process by which snails and their relatives develop their characteristic shape, is the most obvious distinguishing feature of the gastropods. They begin embryonic life as a bilaterally symmetrical larva, but at a particular point in development the hind end twists round as much as 180° and comes to lie near the head end. From this point on, development is asymmetrical, and the internal organs on the undeveloping side are frequently suppressed. The body continues to elongate on its dorsal side, forming a spiral pattern which is reflected in the spiral of the shell enclosing the body. Some gastropods have secondarily lost their shells in adapting to new environments and ways of life.

The polycypods are one of the largest groups of molluscs and include all those with paired lateral shells joined together by a dorsal hinge. These so-called bivalves include the clams, oysters, scallops and mussels. The foot is compressed to form a muscular spade for digging and the head is greatly reduced. The two halves of the shell can enclose the body completely in most species, and two strong muscles, the adductors, can hold it tightly closed against predators. The shell is secreted by the inner lining of tissue, the mantle, and therefore is enlarged as the body of the animal gets bigger. The edges of the mantle around the free margins of the shell are kept closely opposed, forming a cavity which contains the gills. At the posterior end, there are two openings in the mantle, the incurrent and excurrent siphons. These may be extended as a long double tube when the animal is buried in mud and are the means by which a constant stream of water is kept moving over the surface of the gills. The lining of the mantle cavity and the surfaces of the gills are covered with minute cilia whose beating keeps the water flowing in a single direction. A mucous sheath secreted by the gills traps food particles such as algae and bacteria which are in the water, and special tracts of other cilia keep this mucus moving towards the mouth where it is swallowed. There is no radula.

The third group, the cephalopods, is very different from other molluscs. At first sight, squid, octopuses, cuttlefish and their relatives do not appear to have anything which would relate them to the others described here. The foot forms a ring of arms and tentacles about the mouth, and they are active, fast-moving animals. Most have no external shell. They have stout horny jaws which assist the radula in tearing food. Their eyes are prominent and image-forming rather than merely light-sensitive. The most primitive members of this group, such as Nautilus, have a secreted external shell, but in other members of the group, the shell has become modified and reduced in size to form an internal supporting structure. The primitive cephalopods dominated the seas for many millions of years, but dwindled to near-extinction. Those that have survived are highly specialized.

▶ Abalone

A close relative of the limpet, the abalone is a single-shelled mollusc. The body consists mainly of a muscular foot and is fringed with tentacles. Water is drawn in under the shell, passed over the gills to extract oxygen and exhaled through a line of holes along the top of the shell. Abalones eat seaweeds which they find by probing with sensitive tentacles on their heads. They are scraped up and chewed by the rasping action of the radula. They inhabit Pacific and Mediterranean shores.

▼ Cockle

There are almost 200 species of cockle, with a world-wide distribution. The most familiar of these is the so-called edible cockle. It is found from the tidal zone to depths of more than 1300 fathoms, in clean sand, buried below the surface. It is a common animal. In one survey carried out in a favourite habitat, a sheltered bay or estuary, more than 10,000 to the square metre were counted.

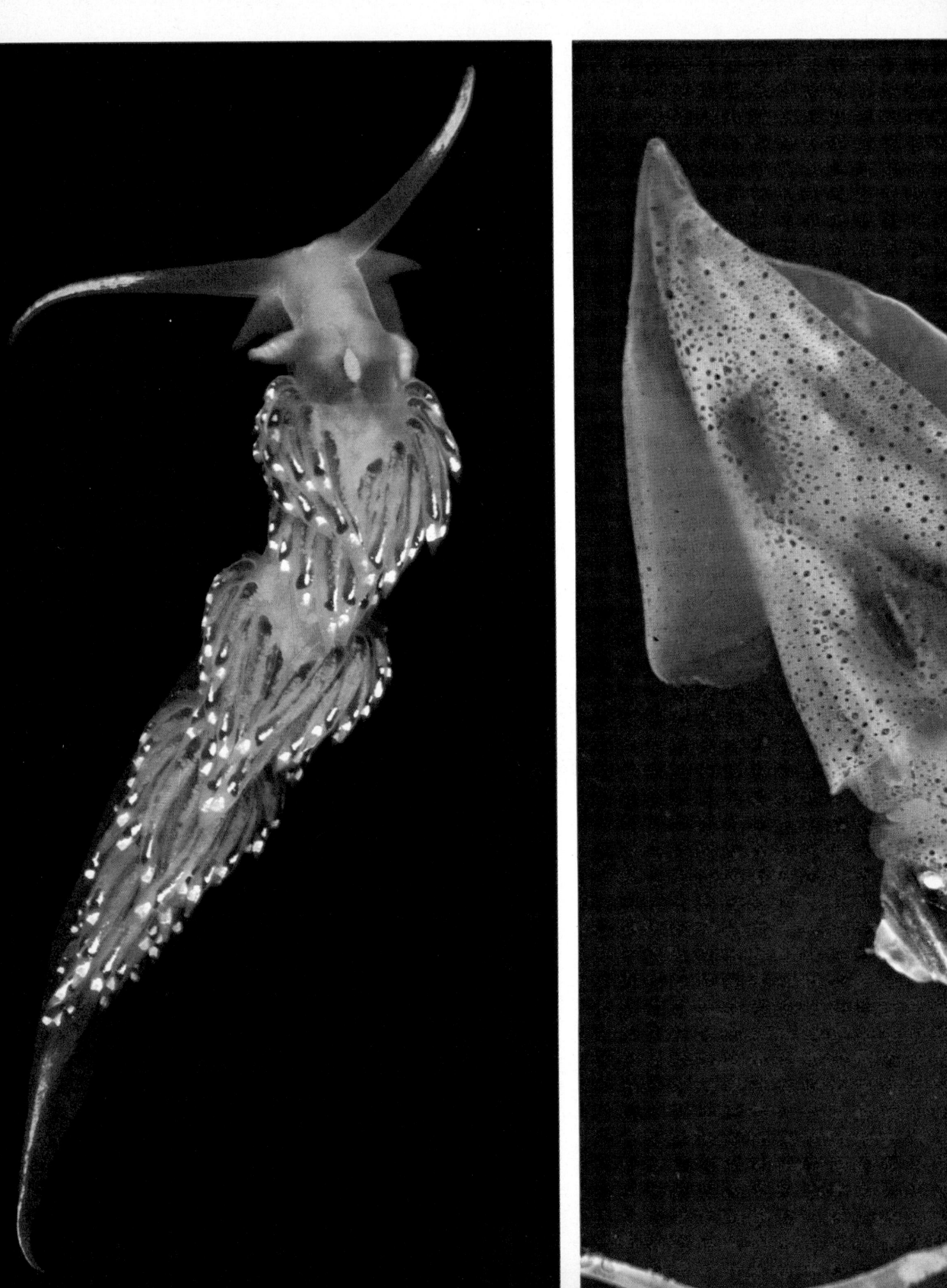

◄◄ Sea Slug *Facelina coronata*

This is another brightly-coloured tropical sea slug. It feeds primarily on sea anemones. The anemone's stinging cells pass through the slug's digestive tract without discharging or causing any damage. They then pass through the tissues of the animal to become lodged in the tentacle-like papillae on the slug's back. The darts collect in special sacs, open to the exterior. The slug has no control over the discharge of the darts as they are not physiologically or anatomically connected to it. But if the sea slug is attacked and one of the papillae is ripped off, the darts will fire. One slug may be damaged by such an attack, but a predator will learn from this experience and in future avoid slugs of the same species. Also its colour may act as warning.

◄ Squid

Squid are cephalopod molluscs. They have two prominent and well-developed eyes which bear a remarkable resemblance to the eyes of mammals. The mantle of the body has two lateral fins and, inside, a supportive structure made of cartilage, the remnant of the mollusc shell. The squid's gills are located beneath the mantle, the layer of tissue which covers the posterior end of the body. It is open at the anterior end and regular muscular pulsations draw water into the cavity, over the gills and expel it again. The water passes through a muscular funnel, and rapid expansion or contraction of this, causes water to move through it with considerable force, allowing the animal to back up or go forward rapidly, an example of animal jet propulsion.

▼ Clam

There is some dispute about which molluscs are clams. Clam is an unspecific word and means different things in different parts of the world. The most obvious parts of the clam are the large hinged shells. They are made of calcium carbonate, deposited by an inner layer of living tissue called the mantle. The mantle is usually extruded along the opening between the shells in a resting, undisturbed animal, and it is a portion of the mantle of the giant clam that is shown in this photograph. The mouth also protects the internal delicate gills. Their surfaces are covered by microscopic extensions of tissue, called cilia, which beat rhythmically, passing a steady stream of water over the gills and aerating them. The gills also collect food.

227

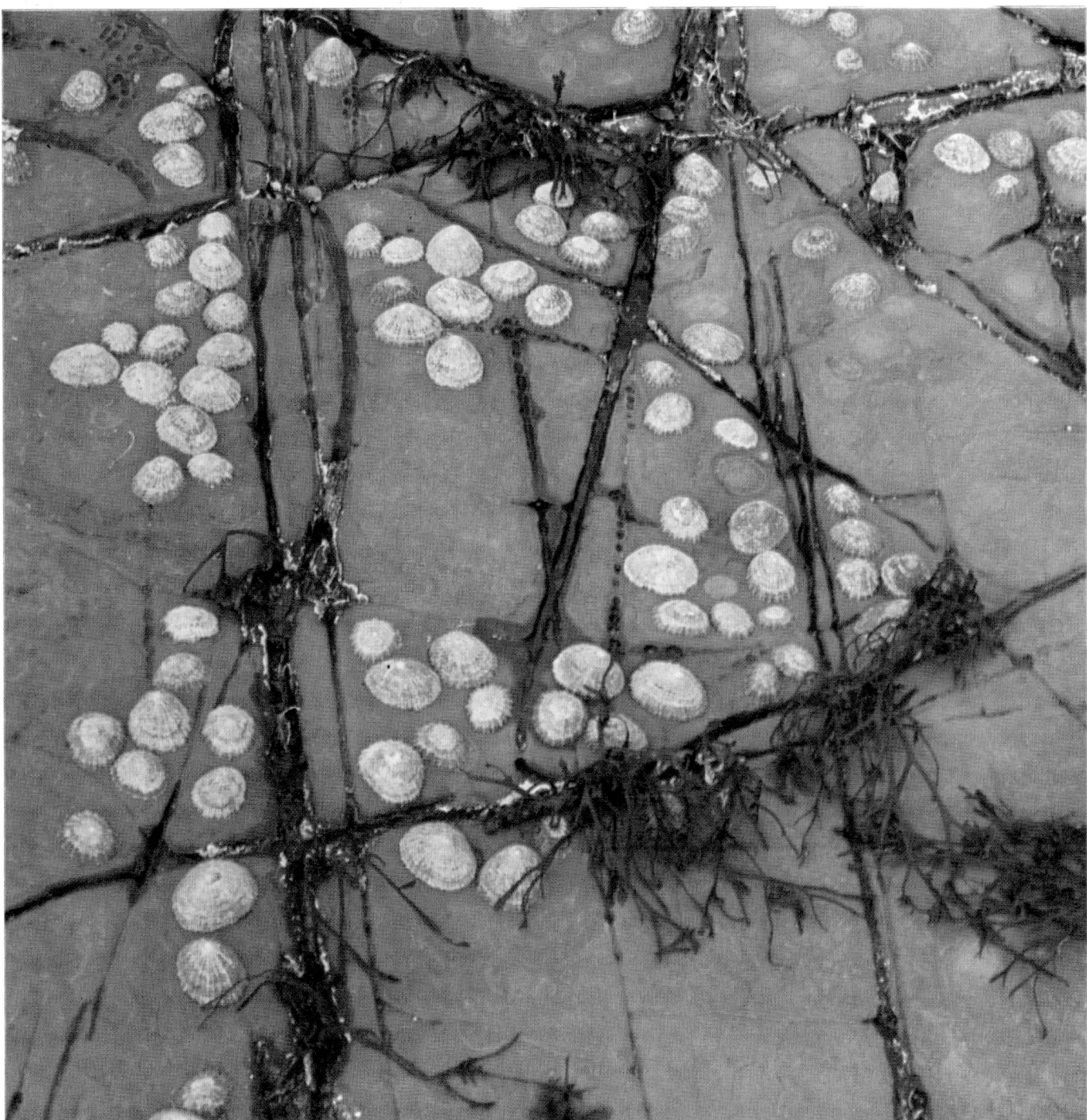

◄ **Octopus**

The name of the octopus partially describes it, for its name means 'eight feet'. The octopus is a relative of the squid and the cuttlefish, but lacks their two long tentacles and the internal remnants of the molluscan shell. The most characteristic feature of the octopus is the long tentacles, joined together by a pliable web at their bases. Each is provided with two more or less parallel rows of suckers which extend the full length of each arm. The body of the animal is short and round, with the mouth at one end, surrounded by the feet. There are more than 135 species of octopus, found in most seas, although they are most common in warm tropical waters. The octopus can disguise itself by changing its colour, extending or curling its tentacles and protruding or concealing its suckers.

▲ **Limpet**

Limpets are relatives of the snails, although at first sight they may not look like it. The name limpet is applied to a variety of different species which have two principal characteristics in common. Their single shell is more or less shaped like half a walnut and they are found clinging tightly to rocks or other surfaces. The limpet shape has evolved separately on several occasions and is particularly well adapted to withstand the hard pounding from heavy surf, to which these animals of the shore line are subject. The body of the limpet is primarily made up of a large, oval foot with a massive flat adhesive surface. There is a distinct head at one end, with prominent tentacles. The shell is lined with a layer of tissue which is called the mantle. The mantle is responsible for depositing the shell.

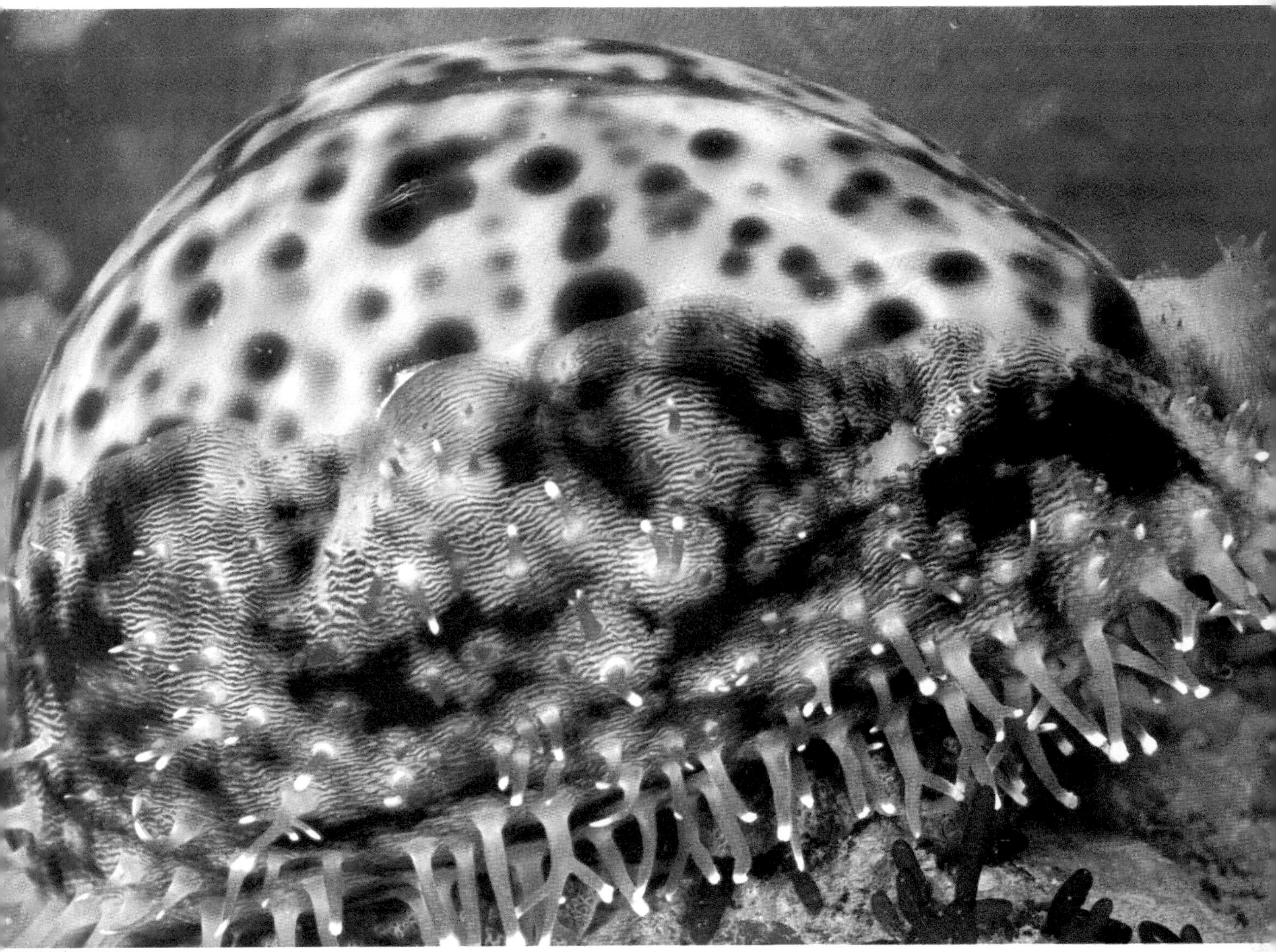

◄◄ Sea Slug

This tropical sea slug looks more like a plant than an animal. This is one of several species which eat plant food and put it to more than one use. It feeds on small green seaweeds or algae. These plants have chloroplasts in their cells. These are green and the principal metabolic organelles in plants. When the plant cells are broken down during the slug's digestive processes, the chloroplasts remain intact, passing into pouches which branch off the gut. There they continue their photosynthetic processes, passing their products to the cells of the slug.

Cowrie

The cowries are relatives of the snails and whelks, but they have lost their usual exterior spiral shape common to this group of molluscs. Shown here are two of the more common species of cowrie, the tiger cowrie (above) and the European cowrie (left). When they are moving about in search of food, the outer mantle usually covers the whole shell of the animal. In the picture above, it is forming only a partial cover. There are two sensory tentacles at the anterior end of the animal, each of which may have a light-sensitive eye at its base. In other species, the mantle itself bears a number of small eyes. In addition to eating seaweed, cowries are carnivorous, eating small sea anemones, sponges, sea squirts, corals, other molluscs and a variety of eggs. The breeding habits of some cowries are rather odd. The female lays a capsule containing several hundred fertilized eggs, frequently in close association with a colony of sea squirts. Several capsules may be laid, fertilization taking place at the time of laying by the female releasing sperms which she has stored in a special vesicle after mating. She then guards the eggs until they hatch, within a few weeks. The young cowries are at first free-swimming larvae, but quite soon the shell begins to develop. At first, it grows with the typical spiral but the last whorl overgrows the entire shell.

Echinoderms

Echinoderms are sedentary or slow-moving inhabitants of the ocean floor. They are conspicuous and common everywhere. They are a group of animals which is greatly diverse in form and function, but there are two characteristics which all five groups of echinoderms possess. All are radially symmetrical, that is, symmetrical about a central point, although this symmetry is not as well-developed as it is in the coelenterates. They are pentamerous, their symmetry being in an arrangement of five or multiples of five. The second common characteristic is the possession of tube-feet, hundreds of tiny organs which are used for movement by acting in concert. They are controlled by a remarkable water vascular system.

There are five separate classes of echinoderms, some more familiar than others. The crinoids are the most primitive. They are attached to the ocean floor by a long stalk and the five arms radiate from its top. The sea lily is a good example. The holothuroids are the sea cucumbers, and their common name is the best short description possible. The asteroids are represented by the common and familiar starfish, while their close relatives the brittlestars and basketstars, the ophiuroids, are also five-rayed, but with much longer and more slender arms and smaller central bodies. The last class is the echinoids, again a familiar group, made up of the spiny sea urchins and sand dollars.

The closest living relatives of the echinoderms are a small and relatively insignificant group of animals called the hemichordates. Echinoderm and hemichordate adults bear few common features, but the details of their embryological development are remarkably similar. In the same way, embryological developments in the hemichordates and chordates, the world's dominant group of animals, show almost identical features in some respects. Thus the echinoderms occupy a unique position, for in or near them are the ancestors of the chordates. The possible origins and relationships between the two is one of the most exciting fields of speculation in the whole of zoology.

► **Featherstar**

At first sight the featherstars do not appear to be typical echinoderms because it is difficult to see their basic five-rayed structure. Each of the five arms of the featherstar splits at its base, producing ten arms, each of which in turn subdivides. This gives the impression of a bundle of arms randomly arranged, as in this brilliant yellow featherstar from the Great Barrier Reef. Featherstars are found in most seas, at a depth of about 100 fathoms, on hard bottoms.

▼ **Sea Cucumber**

There are more than 1000 species of sea cucumbers, the group of echinoderms which look most like plants. The basic five-rayed structure, although present, is not easily seen externally except at the anterior end of the animal. Here the tentacles, not visible in this photograph, are arranged in a circle in multiples of five. Sea cucumbers are found in all seas, in a variety of shapes and sizes to adapt them to a variety of environments.

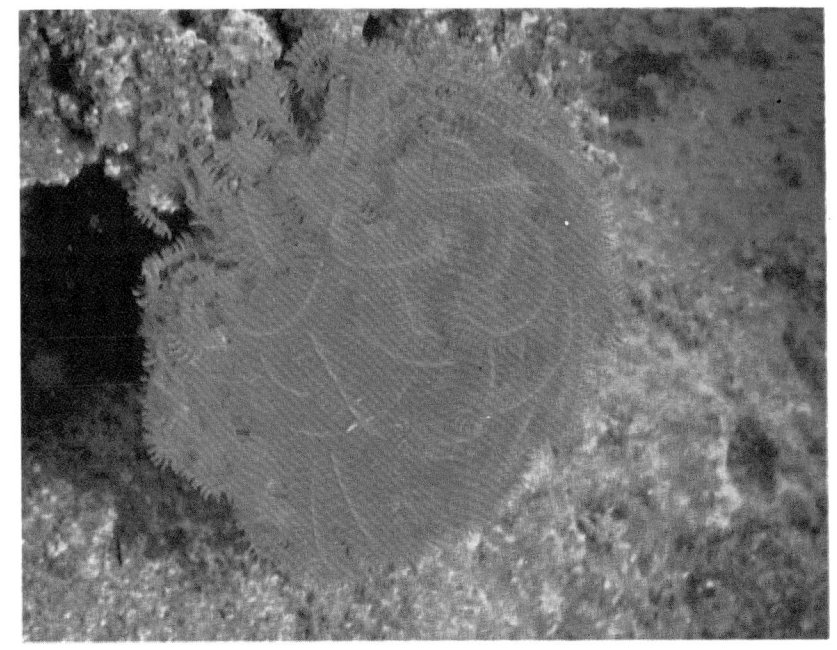

Brittlestar

The brittlestars are echinoderms which are related, although not particularly closely, to the starfish. There are five long arms radiating from a central body. The body is so small in relation to the length of the arms that it is often called a disc. A large brittlestar may have a disc 10 cm in diameter, although the arms have a span of up to 60 cm. The animal is protected by a hard covering of plates and spines on the disc and the arms. The plates on the arm are articulated, and the muscles of the arm are attached to them, giving supple movement which is not dissimilar to that of the flexing of a mammalian tail. There are many different species of brittlestars. There is one species which seems to be found in almost all marine waters. Other species are less widespread but there are brittlestars of one form or another in all the oceans.

Brittlestars take their name from the fact that their arms fall off, or can be pulled off, very easily. This is not a particularly serious problem, for new appendages grow immediately to replace lost ones. Some brittlestars live on the seabed, and prefer a muddy bottom into which they can burrow. Others live amongst the seaweeds or corals. They are not particularly mobile although they can use two or four arms, working in pairs in a sort of rowing action to propel them across the bottom, or allowing them to 'swing' through the seaweed in an acrobatic fashion.

Brittlestars feed either actively or passively. Passive or filter feeding involves the entrapment of small particles of organic matter or of microscopic organisms, in threads of mucus which extend from the arms. Organic debris and micro-organisms become entangled in the mucus, which is continually pulled back towards the animal by the action of thousands of minute cilia which cover the areas. The ciliary action carries the mucus towards the so-called 'tube feet', a characteristic structure in echinoderms. Their action masses the mucus and passes it from one tube foot to the next. This results in the formation of a mucoid ball which is passed down the arm,

234

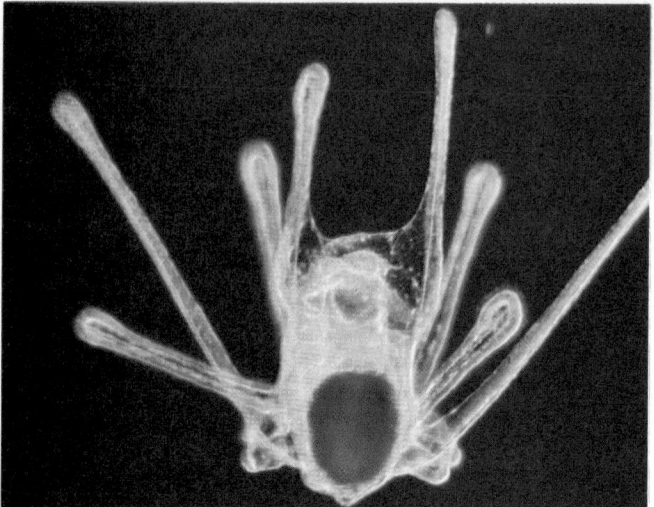
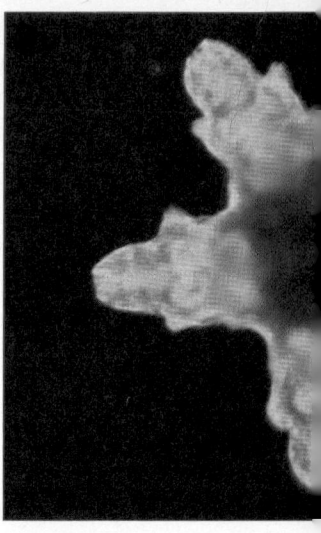

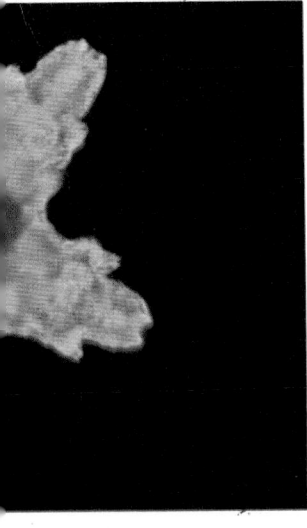

being added to all the time, until it reaches the mouth. Tube feet around the mouth examine the ball, and if it is acceptable, it is ingested. If it is unacceptable it is passed back down the arm and released from the end. The part that the tube feet play in feeding in brittlestars is quite unique, as the tube feet in most echinoderms are concerned with locomotion, and this function, plus their shape, gives them their name.

Breeding is a relatively simple process. In most cases, eggs and sperm are shed into the sea via genital openings at the bases of the arms. Fertilization is therefore external, the fertilized egg developing into a delicate long-armed larva covered with cilia whose action keeps the animal from sinking for quite some time. Two stages of development of the larva are shown at the top. The newly-hatched larva (left) does not bear much resemblance to the adult. By the time the larva has metamorphosed (right), it shows the basic structure of the adult even at this stage, when the tiny brittlestar in this picture has been enlarged to 90 times its actual size. A few species of brittlestar have different breeding habits, in that fertilization is internal, and the larvae are retained inside the body of the female until they have metamorphosed, then being released as miniature adults. Some of these species are hermaphrodites, but self-fertilization is prevented by the fact that the ovaries and testes mature at different times.

◄ Sea Cucumber

This brilliantly-coloured sea cucumber, a tropical species from the Great Barrier Reef, gives a better idea of the general structure of this group. It is more or less U-shaped, with the posterior end of the body slightly upturned. The feathered tentacles are particularly well seen at the anterior end of this specimen. Several species of molluscs use the interior of living sea cucumbers as a home, and some, like one species of snail, have become so adapted to this kind of life that they have become reduced to a bag of reproductive organs and nothing else.

► Featherstar

This photograph gives some idea of the typical posture of most featherstars. The arms radiate out from the central body in a bowl-like formation. Food – organic debris or plankton – drifts into the current to be caught up in the mucoid secretion of the arms.

▼ Starfish *Protoreaster lincki*

The starfish are probably the most commonly known of all the echinoderms. The most familiar types are the typically arrayed five-armed specimens, but the number of arms may vary from four to fifty. Size is also variable, ranging from less than 1 cm in the smallest species to more than 90 cm in the largest. This photograph shows a number of specimens of a common species of the Indian Ocean, together with a hermit crab and another common echinoderm, a sea urchin.

237

Cartilaginous fish

It is important, at the outset, to define just what is meant by the word fish. Some people seem to apply it to any aquatic vertebrates, and some even apply it to aquatic mammals, in expressions like 'whale fish'. Others refer to various aquatic invertebrates – crustaceans, molluscs and echinoderms – as fish. 'Fish' is not a scientific term, but strictly speaking it should be used to describe aquatic chordates which have fins and gills.

Using this definition, there are three living groups of fish and a good fossil record to determine their relationships with their extinct ancestors and relatives. First, what is the difference between a chordate and a vertebrate? Chordates are animals which have, at some stage in their lives, a long dorsal rod of tissue called the notochord, which serves to stiffen and support the body, gill pouches in the pharynx and a single, dorsal tubular nerve cord. Vertebrates have a cranium encasing the brain and separate vertebrae in the backbone, replacing in the adult the more primitive notochord. Thus all vertebrates are chordates, but not all chordates are vertebrates.

The origins of the chordates, which gave rise to the vertebrates, are not clear, but as pointed out in the section on echinoderms, they probably came from a relatively obscure and now extinct echinoderm-like ancestor. The record of the subsequent evolution of the vertebrates is now reasonably complete. The most primitive were small, fresh-water, bottom-feeding animals called the agnathans, meaning jawless. They were generally fish-like in appearance, had a thick armour of bony plates and scales below the surface of the skin and a skeleton made of cartilage. They had median but no paired fins, a single median nostril, a pair of lateral eyes and a third light-sensitive, but probably not image-forming eye called the pineal eye. They had several to many pairs of gills, and in most species there was a separate gill slit, or opening to the exterior, for each gill. The only living members of the agnathans are the lampreys and the hagfish, known as cyclostomes because of their circular, jawless mouths. They are primitive vertebrates in many respects, but they are modified and degenerate offshoots.

The next most primitive group of living vertebrates are the skates and rays, the cartilaginous or elasmobranch fishes. They represent a significant advance over the agnathans. They have jaws and paired fins, well-developed scales, are hydrodynamically designed to fit their habitat and behaviour, and a reduced number of gills and gill slits. Like their more primitive ancestors, they evolved in fresh water and then invaded the sea, so that most living elasmobranchs are marine. The skeleton of the sharks, dogfishes, skates and rays is primitive in many respects, but the fact that it is made of cartilage and not of bone is not a primitive characteristic. The elasmobranchs have evolved from ancestors with bony skeletons, and the presence of the cartilaginous skeleton is a retrograde condition.

Eagle Ray
This eagle ray, one of several species, derives its name from the fact that its shape is slightly different from that of other rays. Unlike other rays and their relatives the skates, the wing-like pectoral fins do not extend as far forward as the head. The snout and head are left clear, giving the general impression of the shape of an eagle in flight, rather well shown in this picture. There is a long tail, armed with a poison spine, and like the other rays and skates, the mouth and the gill openings are on the underside of the body. Eagle rays are primarily tropical, but they are caught occasionally in temperate waters. Unlike their relatives, the eagle rays do not spend most of the time on or near the bottom of the sea, but seem to prefer to spend their time swimming gracefully through the upper layers of the sea, occasionally breaking the surface of the ocean and skimming short distances over the surface before descending again.

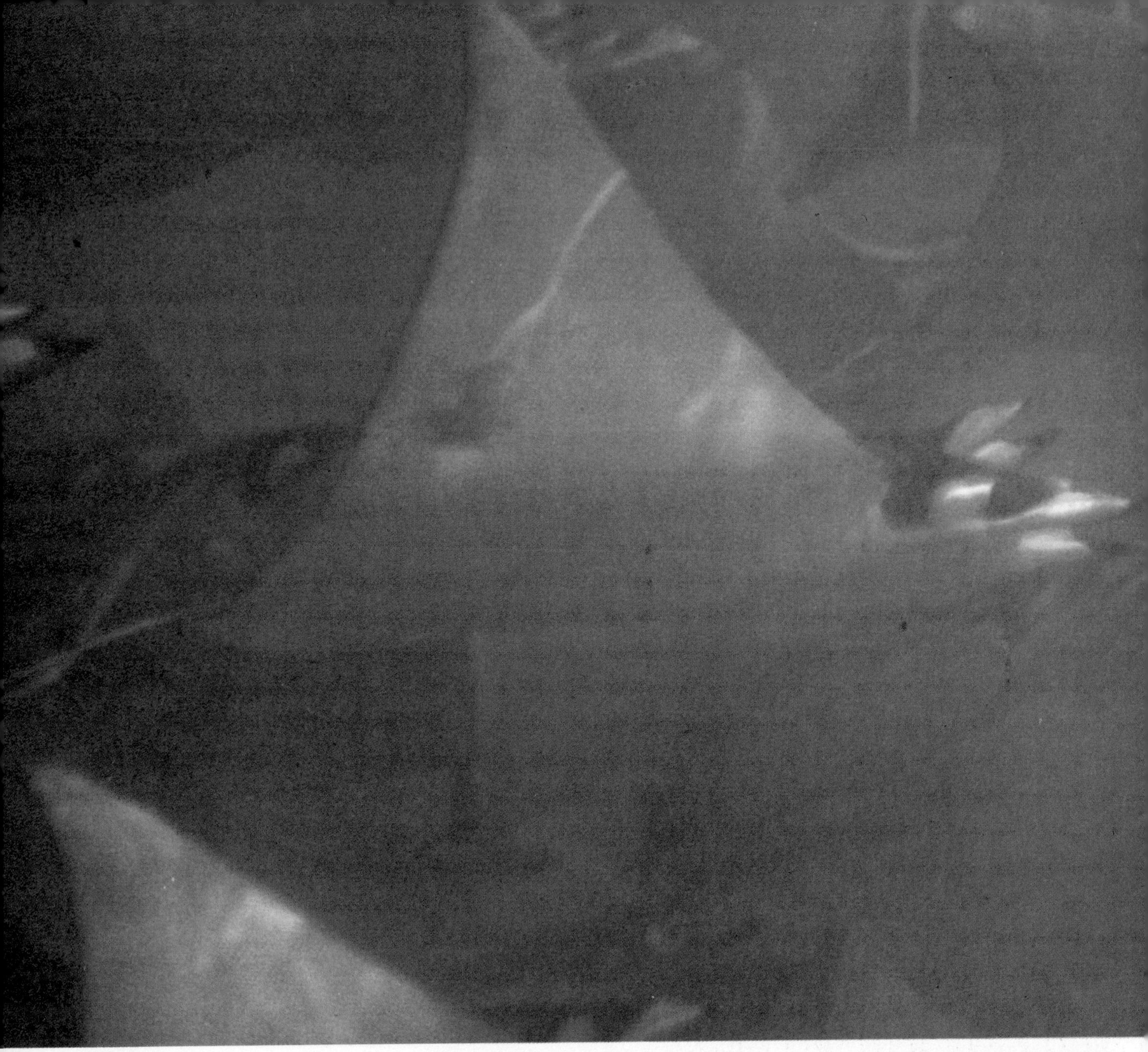

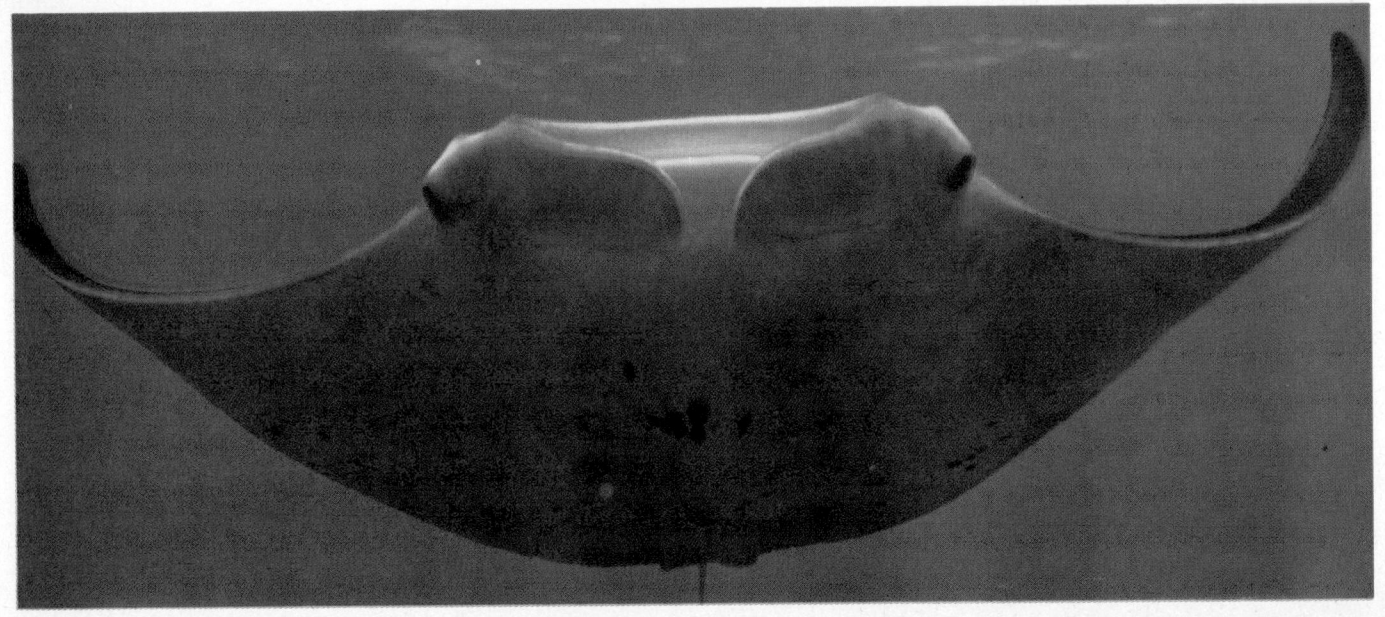

Devil Fish

*There are fewer than 20 species of devil fish, probably
better known by their more common, if slightly
inaccurate name of manta. The mantas are some of the
largest rays and may have a 'wing span' of more than
20 m and can weigh as much as a rhinoceros. The typical
ray shape of a flattened body, wing-like pectoral fins
and a long tail, in some equipped with a poisonous spine,
is slightly altered in the devil fish by having an anterior,
rather than underside mouth, which is relatively large
(left). The 'horns' at either side of the mouth are formed
from extensions of the pectoral fins. Like the eagle rays,
mantas spend a good deal of their time in the upper
waters of tropical and sub-tropical seas. Their bat-like
flight has prompted a series of common names referring
to this characteristic movement – sea bat, batfish and
vampire ray. Unlike some species, these rays may form
schools (above), although they are more usually
solitary or in pairs.*

*Devil fish are filter feeders, in spite of their large size
and predator-like appearance. They swim through water*
*rich in planktonic food, and a special filtering system
removes the food from the water before it passes over the
gills to the exterior, and channels the food into the
oesophagus. Because of their size and appearance, devil
fish have a bad reputation, but this is probably for the
most part untrue. A leaping devil fish can smash a small
boat if it lands on it, or tow a boat if it has been
harpooned, but these are the results of either deliberate
interference with the animal or an inadvertent move on
its part. Because of their size, they have been difficult
to study, but evidence from skindivers indicates that
devil fish are for the most part gentle, inquisitive
creatures, with no malicious streak in them. This is
probably illustrated by the fact that divers can hold on to a
swimming devil fish and hitch a ride with it for some
considerable distance.*

*During mating, the male and female swim together
and seem to embrace each other with their large
pectoral fins. Usually, only one young is born at a
time but it is quite large and very well-developed
at birth.*

◄ **Sand Shark**

The sharks are probably the most feared of all the marine animals, but only partly for good reason. They are for the most part active predators and they are certainly known to attack man. This photograph illustrates particularly well the structure of the shark's mouth. It is provided with sharp teeth, and the position of the large mouth makes it necessary for the animal to roll over on its side or back to make an attack. Attached to the underside of this shark is a remora, a bony fish whose first dorsal fin has become adapted into a powerful sucker.

▼ **Skate**

There are more than 100 species of skate, one of the more familiar cartilaginous fish because of its commercial importance. Skates are amongst the smallest group of the rays but externally the close relationship is obvious. Skates are for the most part bottom dwellers, feeding on crustaceans and small fish, which they capture with a pouncing movement, so smothering them with their large pectoral fins and then attacking with the mouth. Like some other marine animals, skates have organs in their tails, producing an electrical charge of about four volts.

Bony fish

As mentioned in the previous section, the sharks and similar fish were fresh water animals in the first instance, as were the first vertebrates. They subsequently developed the physiological mechanisms which allowed them to enter and exploit salt water environments, then under the domination of the giant cephalopod molluscs. While this was going on, and the elasmobranchs or cartilaginous fish, were establishing themselves as the dominant form of marine animal life, another offshoot of the primitive vertebrates, the bony fish, was developing rapidly and becoming dominant in fresh water.

The bony fish resemble their cartilaginous relatives in being evolutionarily advanced fishes. They have efficient paired appendages, the pectoral and pelvic fins, for balance and fine movement, powerful tail and body muscles for propulsion, and strong jaws with teeth. Most obvious, perhaps, is the presence of a bony skeleton which has replaced the more primitive cartilage. In embryological development, the skeleton is first laid down as cartilage and is then eroded and replaced by bone.

Two other characteristics indicate the evolutionary superiority of the bony fish. The first is the presence of the operculum, a bony flap which covers the gills and reduces the number of external openings of the gill chambers to one on each side. The second is the presence of the swim bladder, a usually single, blind, median, gas-filled sac which is an outpouching of the pharynx. For many years it was believed that the swim bladder was the evolutionary precursor of the lungs of the higher vertebrates, but the reverse is probably more likely. Lungs probably developed in early freshwater fish to enable them to survive in stagnant water or during drought, as in the living lungfish of Africa, South America and Australia. Thus the freshwater bony fishes which invaded the sea probably had at least rudimentary lungs, which had enabled them to survive the unreliable freshwater conditions which had occurred earlier. Those freshwater fish that did not have lungs either became extinct or, like the sharks, moved to the sea. Life in the sea presents its problems, but not droughts. Therefore the lungs are relatively useless for respiration, but have become adapted, as swim bladders, to function as hydrostatic organs. By secreting gas into, or absorbing gas from, the swim bladder, the fish can adjust its specific gravity, and thus is able to maintain itself at varying depths in the water with minimal effort.

Scorpionfish

The 300 or so species of scorpionfish include some of the most beautiful as well as some of the ugliest bony fish. Their special appearance is due to two particular features. The head is massive, with the eyes high and the mouth wide and downward-sloping. The fins are divided into narrow strips supported by spines. There are also numerous poison spines scattered amongst the fins. The body is usually covered with brightly coloured stripes. Most scorpionfish are found in temperate seas, although there are a few natives of the tropics.

► **Scad** *Trachurus trachurus*
The scads were originally known as horse mackerel, but although the two do look alike, there are several differences which warrant the separation of the two species. There are three species of scad and the one shown here is the common scad, a native of the North Atlantic. It is most common in European waters but is also found off the coast of North America. The other two species of the genus Trachurus are also found in the North Atlantic as well as in the Mediterranean and Black Seas.

The scad are identified by their torpedo-like shape. They have a prominent lateral line, well shown in this photograph. The lateral line is a sensory organ found in all the fish-like aquatic vertebrates and runs from the head to the tail. It is provided with minute sensory cells which are sensitive to low frequency vibrations, movement and perhaps to pressure changes in the water, telling the fish what depth it is at. In the scad, the lateral line is covered by a series of small bony plates and is one of the fish's distinguishing characteristics.

Scad spawn in the North Atlantic during the summer months. The eggs have a relatively high acid content, so that they float near the surface of the water until hatching. The young larva is small and finless. Young scad feed on diatoms and other plankton such as larvae of fish and crustaceans, and eggs of a variety of types of animals. As they grow older, they form into schools or shoals. They continue to be carnivorous, feeding on other fishes such an anchovies, herrings, sprats and pilchards. They will also eat molluscs such as squid, and crustaceans.

Scad have little or no economic importance as a food fish. Large commercial fisheries do exist off the coats of Portugal and Spain, where they are eaten, but for the most part they are used to make fish meal.

There is no definite information as to the predators of the scad, but it seems most likely that these would include larger, carnivorous, bony fish as well as porpoises and dolphins. Scad have developed the habit of seeking shelter among the tentacles of certain jellyfish. Presumably they themselves are immune to the sting of the nematocysts, and it has been reported, although not confirmed, that they eat the tentacles and even the eggs. They also prey on other aquatic animals which seek refuge among the jellyfish's tentacles.

246

▶▶ **Stonefish**

The stonefish is almost certainly the ugliest of all the bony fish, and is doubtless the most poisonous. The head is broad and flat, rather like that of the toadfish, and the body tapers rapidly to the tail. The mouth is large and wide. The skin is scaleless and is covered with warts and bumps which are in turn covered by a layer of mucus. The stonefish is able to vary its coloration in relation to its background, so that it can be mud-coloured, stone-coloured, or even like a variety of types of seaweed. There are three species of stonefish, extending from the Red Sea to East Africa, India and the coasts of Australia, and they are found in shallow seas on rocky, coral-covered, or muddy bottoms.

The stonefish's principal defence, other than its superb camouflage, is its powerful, even lethal, poison. The fin rays, bony extensions which support the fins, have become adapted as poison spines. There are 13 of these spines along the dorsal fin, three on the anal fin, and one on each of the pelvic fins, giving the stonefish all-round protection. Each spine has two poison sacs near its tip. The spine is covered with a sheath of skin, but pressure on the tip of the spine causes the sheath to slide back, exposing the needle-sharp shaft. This has two grooves in it leading to the poison sacs. The poison flows along these grooves and into the wound. When disturbed, the stonefish lies completely still, the only movement being the almost undetectable erection of the spines. Even when caught the stonefish is still dangerous, for it can stay alive for up to 10 hours out of water, and even when dead can still inflict a serious injury, as the method of delivering the poison is purely passive. Although some reports of swimmers tangling with stonefish have noted little or no damage, there are well-authenticated cases of agonizing pain, and even death, from a large dose of stonefish poison. Part of the variation from person to person in reacting to the poison is probably explained by the likelihood that some people are more sensitive to the poison than others, in the same way that different people have varying degrees of reaction to mosquito bites. The stonefish is carnivorous, lying quietly on the bottom until suitable prey swims past, then grabbing it with its large mouth. The stonefish's move is so fast the prey just seems to disappear. Surprisingly, perhaps, the stonefish does have a few predators including sharks.

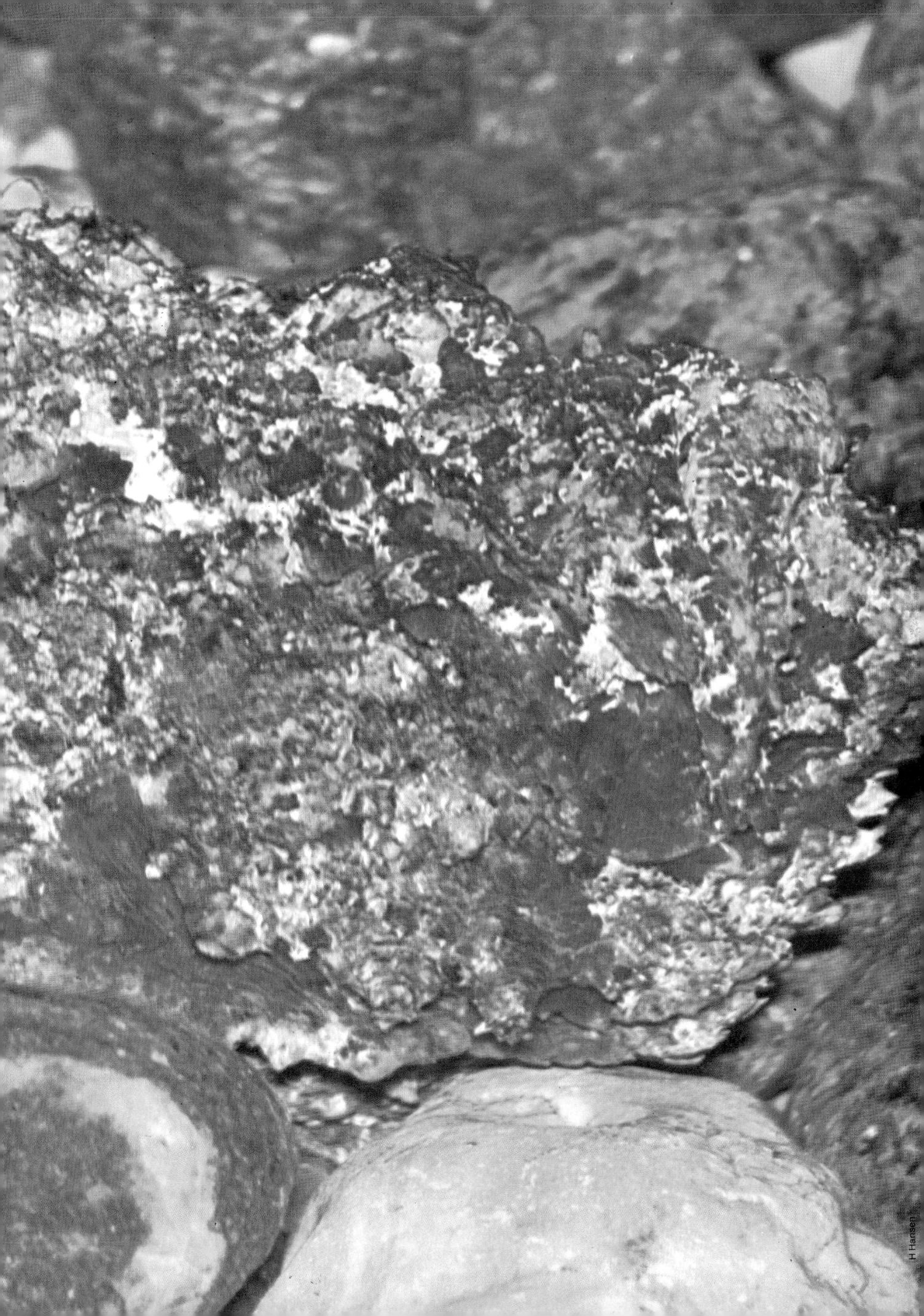

▶ Mullet

This is a school of young mullet, or fry, stranded in a pool. Pools are frequently inhabited by young mullet, who presumably are protected from predators in this way for at least part of every day. Mullet spawn in shallow water, clustering together tightly. The eggs hatch quickly and the young fry begin to feed. They live together in shoals during the day, dispersing at night with each fish going to its own spot on the ocean floor. Any disturbance will cause them to form quickly into the shoal once more. They are primarily filter feeders, sucking up sand and mud into their mouths and extracting the organic material from it by a sophisticated filtering system. They spit out the inedible portion. They also feed at the surface, again using a filtering system, and are particularly numerous when the tide is running out, picking up small animals and particles of food as they are swept out to sea. As well as having taste buds inside their mouths, they have them scattered over the outer surface, especially the head. They also eat small crustaceans and carrion and will scrape small filaments of algae off stones, piers and other solid objects under water. Mullet are tropical and sub-tropical fish moving into temperate waters in the summer months.

Gurnard

There is one particular feature which separates the gurnards, or sea robins (left and below), from all other fish. This is the two or three rays of the wing-like pectoral fins which are separated out to form tentacle-like feelers. These serve two purposes, the first of which is locomotion. The gurnards are bottom dwellers for the most part, and their movement across the ocean floor is almost entirely due to these feelers which they use in the same manner as a punt pole. They lift the body of the fish off the floor, moving it forwards, backwards or sideways as it progresses across the rocky or muddy bottom. The pectoral fins themselves, as well as the anteriorly-placed pelvic fins, are held close to the body during this kind of movement and are used mainly for balancing.

The other use of these false legs is sensory. Most fish have in their mouths, on their lips, tongues and heads, cells which are sensitive to chemical stimuli, used for discovering food sources. However, in these bottom-dwelling gurnards, these sensory cells are concentrated on the feelers, which they use to detect their prey, mostly small crustaceans, small fish and sand eels.

Gurnards are also notable for their habit of grunting. When they are removed from the water, distinct grunting noises can be heard, some of them most vocal during the breeding season. The noise is made by the contraction of the swim-bladder, by special muscles.

▲ Moray Eels

There are about 120 species of moray eels, of which six are shown above. They are inhabitants of warm seas in tropical and sub-tropical areas, living on or near coral reefs at depths of up to 15 fathoms. Although the eels are true bony fish, or teleosts, they have several interesting anatomical modifications which separate them from most other teleosts. The skin is thick and the scales have been lost. In some species, the dorsal fin has become greatly enlarged, and may start just behind the head, as can be seen in some of these photographs. In other species the dorsal, tail and anal fins may be absent. In all eels, the pectoral and pelvic fins are either greatly reduced or absent, as in the morays. These modifications to some fins and loss of others, combined with the greatly elongated jaws, give the eels in general, and the morays in particular, a marked, snake-like appearance.

It is probably their appearance, rather than their actual behaviour, which has given the morays a fierce and sinister reputation. Like most snakes, they do not appear to deserve it. Stories are told of them viciously attacking bathers and divers, even to the extent of holding a man under water until he drowns, but none of these tales has ever been substantiated. Divers are frequently searching for the same sort of food as morays – crustaceans and molluscs – and a man probing about a coral reef may inadvertently disturb a moray, lurking unobtrusively in its lair, or may find himself in direct competition for the same prey. A man's hand may even be mistaken for suitable moray food – a small octopus for example. Under these circumstances, or when a moray is molested or attacked, some sort of encounter is almost inevitable, and some divers have sustained severe wounds.

Morays are carnivorous, and will eat anything, dead or alive, that they can find. They bolt their food whole, because they need to keep a continuous flow of water over their gills.

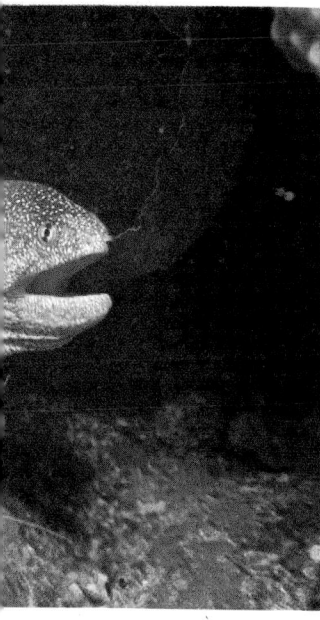

▶ **Triggerfish**

Triggerfish are natives of tropical waters, for only in the warm seas are such bright or even gaudy colours and patterns found. Triggerfish are found in coastal waters of tropical and sub-tropical seas around the world. Three species are shown here: at the top, the Pacific triggerfish, the spotted triggerfish, in the middle, and the queen triggerfish, an inhabitant of the Atlantic and Indian oceans.

The body of the triggerfish is covered with small bony plates, each of which bears one or more small spines. The pectoral fins are reduced in size, and the pelvic fins are little more than short, sharp spines. Of the two dorsal fins, the first is spinous and more or less rigid, although it can be raised or lowered. Thus, with these modifications to the body and some of the fins, the triggerfish must depend on its second dorsal, tail and anal fins for propulsion through the water. Therefore, as might be expected, it is not a strong swimmer and does not manoeuvre easily. The first dorsal fin has an interesting adaptation. There are two or more strong spines, and the first of these cannot be depressed when erect unless the second is also lowered. This is accomplished by a simple but ingenious interlocking device. Thus, when threatened, the triggerfish can take refuge in the rocks or corals, and with its dorsal and pelvic spines, make it impossible for a predator to pull it out.

Squirrelfish

There are about 70 species of squirrelfish, natives of tropical seas and most frequently found in shallow water. These two photographs are of a small shoal of Myripristis (below), found in the Red Sea, and of a single Hawaiian striped squirrelfish, Holocentrus zantherythrus. They get their name from a rather tenuous resemblance to squirrels, in that they are usually red and have large eyes. Squirrelfish are nocturnal. During the daylight hours they conceal themselves in cracks or openings in coral or other appropriate hiding places. They are highly territorial animals, defending their area against intruders and sending loud warning calls to advertise their presence and attract mates. The call is produced by the vibration of muscles which are attached to the swim bladder, which acts as a resonator to amplify the sound, frequently loud enough to be heard above the water.

Because of their nocturnal habits they have good eyesight and protective spines, seen particularly well in the photograph of the striped squirrelfish. The adults have few enemies, except for commercial fisheries in some areas, particularly in the Pacific. They are highly predatory fish, and this is probably one of the reasons for their pronounced territorial behaviour, so that they are spread out over the area they inhabit and do not compete for food.

Their only vulnerable stage is when they are young. The newly-hatched larvae swim to the surface and become part of the planktonic fauna. At first they have no spines, and are at the mercy of the currents. Many larvae and young squirrelfish are eaten by predators, particularly tuna. The squirrelfish are thought to be among the more primitive teleost fish, because they have retained several anatomical features of their ancestors. Because they conceal themselves during the day and because of their protective spines and nocturnal habits, the adults have few enemies.

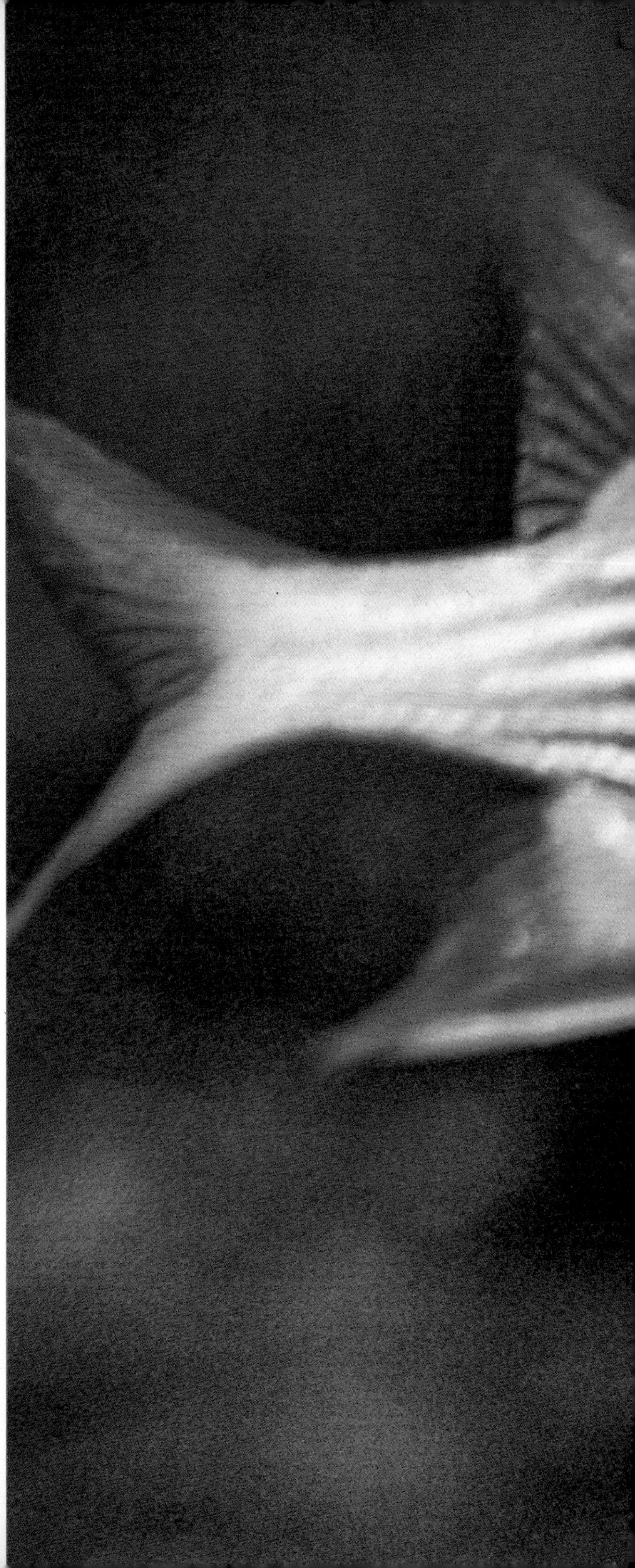

◀ **Archerfish**

The derivation of the name archerfish is described better in this superb photograph than any written description could possibly attempt. The jet of water droplets is achieved by a rapid closing of the gill covers, forcing the water in the mouth along a groove in the roof and ejecting it through a narrow opening of the mouth. The method by which the archerfish takes aim and hits its target accurately is quite complex. The accuracy of the archerfish's markmanship has been shown to increase with practice and with age.

There are five species of archerfish, found in the oceans of south Asia, from India through to Australia. They are members of the perch family. They are not entirely marine animals, as they inhabit both brackish and fresh water, but as the adults return to the sea to breed, it is possible they were once confined to salt water.

▼ **Toadfish**

The toadfish is so-called because of the shape of its head and the loud foghorn-like noise that it makes. There are more than 30 species of toadfish, most of which are found in shallow tropical and temperate seas, although some are found in brackish waters and a few are found in fresh water. They spend most of their time lying on the bottom, and may move so infrequently that they become covered with silt and debris which settles on the bottom. They are territorial animals, vigorously defending their own small areas against other toadfish or any other animal which ventures near them. The loud grunting noise they make is probably part of this territorial behaviour, as well as functioning in the breeding season to attract a mate. Eggs are laid in a sort of nest in any covered or protected areas, and are aerated and protected by the male toadfish until they hatch.

▶ Trunkfish

The trunkfish is one of the strangest of all the bony fish.
Some people have described it as rather turtle-like, not
only because of its appearance, but also because of its
structure. Trunkfish, also known as boxfish or cofferfish,
have their bodies enclosed in a protective covering of
closely adjoining six-sided bony plates, leaving only the
fins and the tail unprotected. The body has become
foreshortened, the head blending into the trunk, the
mouth reduced to a small round hole, armed with
powerful crushing teeth. The pelvic fins have disappeared,
and the number of vertebrae in the backbone considerably
reduced. The long anterior and posterior spines provide
additional protection as do the horny ridges along the
back of the animal. The gill slits, the openings of the gill
chambers to the exterior, have decreased in size, thus
reducing vulnerability. The box enclosing the animal is so
rigid that only the tail, fins and lips are capable of
independent movement. Trunkfish are found at, or near,
the bottom of warm marine waters, especially in the
tropics, in all parts of the world.

It is obvious from this photograph that the trunkfish is
anything but streamlined. The large blunt head produces
considerable resistance as the fish moves through the
water. Combined with this is the inflexibility of the body.
Most fish use side to side movements of the body, by the
flexion of the powerful body muscles, to provide much of
the thrust for forward movement, but because the
trunkfish lacks this flexibility, it has to rely mostly on the
movement of its tail to propel it. This results in rapid and
seemingly laboured tail movements, so that the
trunkfish has to make considerable effort to cover a
relatively short distance at a very slow rate.

Fortunately for the trunkfish, however, rapid
movement is not necessary when it is so heavily
armour plated. It has other protective mechanisms as
well. One of these is the ability to undergo rapid and
dramatic colour changes. One species found off the coast
of America is green with blue spots and lines, but it can
change this to a yellow background with blue spots or
brown with light blue, or even at certain times, become
almost completely white. In other species, there is
considerable sexual dimorphism, that is, the males and
females of the same species may have different colours or
markings, in the same way as many species of song birds.
The young fish of these species may also be differently
coloured again, attaining the adult coloration with sexual
maturity.

This bright, often gaudy colouring is probably a
warning to other fish to steer clear of trunkfish. Not only
are they almost invulnerable, but studies in laboratories
indicate that some species of trunkfish release into the
water a kind of poison, toxic to other fish. The source and
nature of this poison is not known, but other fish sharing
an aquarium with a trunkfish soon become distressed,
come to the surface to gulp air, and die soon afterwards.
The poison is a persistent one, remaining in the tank
after the trunkfish has been removed. Other
trunkfish are not affected by this poison and some robust
fish, such as the moray eel, are also immune.

They eat small crustaceans, molluscs, worms and
other small invertebrates, which some species catch
by blowing jets of water into the sand on the seabed.
They also eat the polyps of corals which are very common
in the tropical waters they inhabit.

Scat *Scatophagus argus*
*The number of species of scat is not known exactly.
There are certainly three, and probably not more than
six, but it is difficult to tell, because all the features used
to distinguish them are so variable. The background
colour of the scat varies through greenish, bluish to
brownish shades of silver, and spots may be scattered, in
rows, or run together in lines or incomplete bars. The
colour of the spots or lines is also variable. There is some
evidence to suggest that these colour changes are related
to age. Scats are found in the oceans south of Asia, from
the east coast of India to New Zealand, in relatively
shallow coastal waters. One isolated species, the African
scat, is found along the coasts of East Africa and
Malagasy.*

Scat

Scat are commonly thought of as inhabitants of marine coastal waters, but in fact they spend much of their time in estuaries and in fresh water. The young especially, frequent waters of less salinity than the sea, and remain there until they are fully grown, returning to the sea to breed. Whether or not they were originally marine fish which were able to adapt physiologically to a partly fresh water existence, or whether they were fresh water inhabitants who were able to invade the sea, is hard to say, although it seems likely that the former is the more logical suggestion. Little is known of their breeding habits, for they have not been studied to any great extent in their natural environment. Although they make good aquarium fishes, they have been bred in captivity only once, and the young did not survive. Eggs are laid in rock crevices and other protected places, and it is thought that at sea they probably lay their eggs in coral reefs. Both male and female guard the eggs and young larvae, who for a short period in their lives have a heavily-plated head and neck, but this protective covering is lost as the fish matures.

They tend to browse when feeding, and appear to eat almost anything – plants, dead animals, mud, decomposing refuse and raw sewage which is dumped untreated into the sea over much of their range. Little is known about their natural enemies, but presumably larger carnivorous fish take their toll.

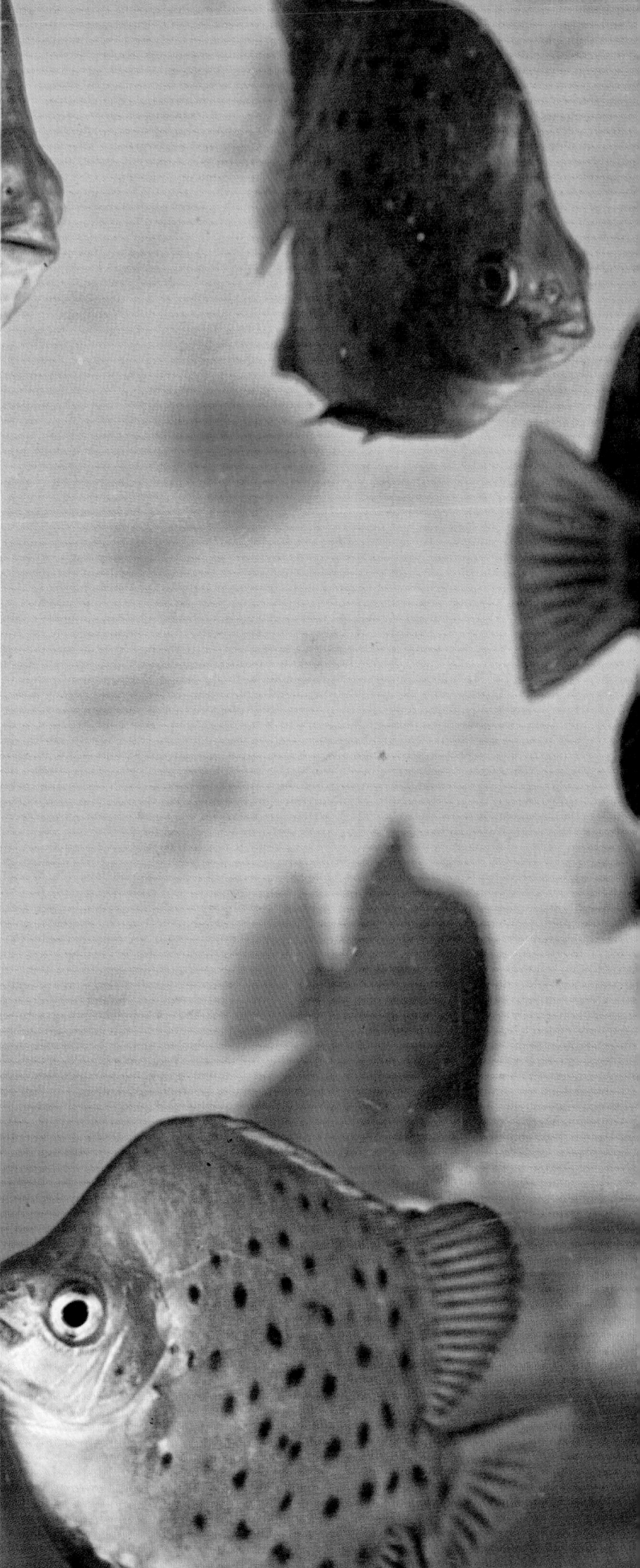

Flounder *Platichthys flesus*

*The flounder (above and far right) is one of the European
flatfish. It is a bottom dweller, found in shallow seas.*

*A newly-hatched flounder looks like any other young
fish. Its body is rounded, the fins are normally
distributed, and there is an eye on each side of the head.
But during the next few months a remarkable series of
changes takes place. The jaws become twisted, and the
dorsal and anal fins, normally relatively small and
concerned solely with balancing, begin to grow and
finally come to extend the full length of either side of the
body. The pectoral and pelvic fins become reduced in
size, and the extended dorsal and anal fins become the
principal swimming organs. The body becomes flattened
and the flounder comes to lie on one side, with what was
the right eye uppermost. Most amazing of all, the left eye
migrates around, over what was the top of the skull, until
it comes to lie alongside the right eye. This remarkable
series of changes, which also occurs in the plaice, the dab
and the turbot, adapts the flounder for its bottom
dwelling habit, giving it the overall outward shape of the
quite unrelated skates and rays, and a definitely lopsided
physiognomy. Like many bottom dwellers, flounders are
able to change colour to some degree to match the bottom
on which they lie. They also are able to spend a great
deal of time in fresh water, being able to cope with the
problems of decreased salinity, but always return to the
sea to spawn. They migrate to particular spawning
grounds and do not feed on the journey.*

*Young flounders eat small molluscs but the adult
fish have a varied diet, changing as they move from
the sea to fresh water. The teeth in the lower jaw
are well developed but those of the upper jaw are
very feeble. However, they have crushing teeth
located in the throat.*

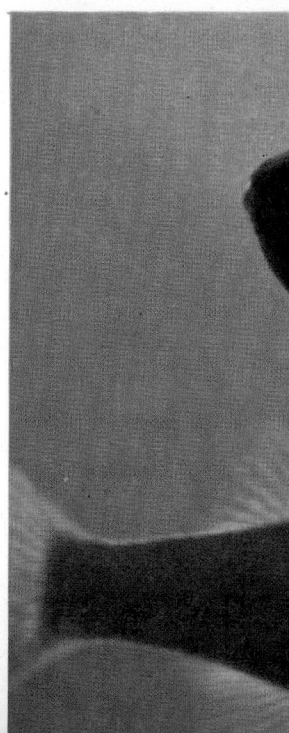

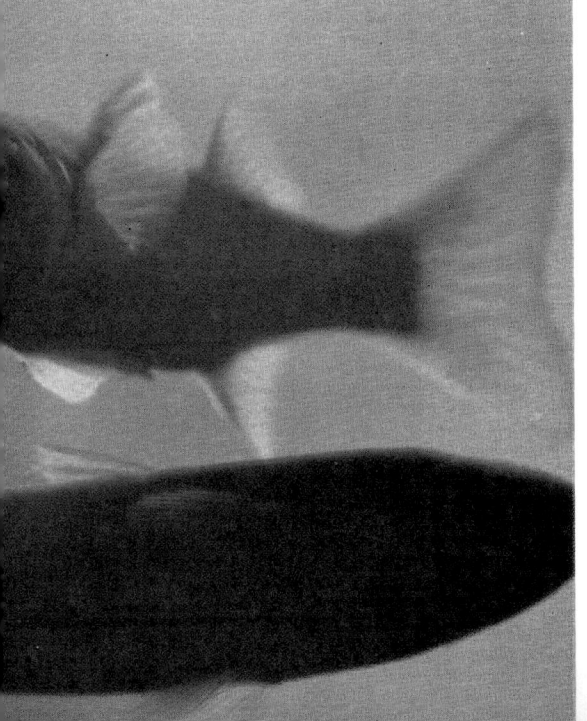

◄ **Mullet**

Mullet are able to move from salt water to fresh water via estuaries, by a physiological mechanism, not completely understood or explained, which allows them to adjust to varying degrees of salinity. This ability is sometimes used for commercial exploitation. Barriers are built across the tidal zones of rivers and lagoons which allow the young mullet to enter but not to escape. In these enclosures, they are farmed and harvested when fully grown. They are difficult to catch, for their delicate mouths tear easily when hooked. Their remarkable jumping ability, combined with the fact that when one mullet jumps all the rest in the shoal follow suit, makes even netting them difficult. Individual mullets in a shoal are sometimes seen swimming upside down.

Blenny

The blennies (left and right) are numerous fish living in shallow tropical and temperate seas. The number of species is not certain but there are at least 20 different families of blennies and each of these has a number of species. They are small fish with elongated bodies. Blennies frequently get caught by the ebb-tide, but are able to survive these periods out of water by sheltering in damp places under stones or seaweeds. Some species are even capable of withstanding long periods out of water, and have been known to climb out of the water voluntarily to sit on rocks or other structures. They are able to extract atmospheric oxygen from the air because of an enlarged thyroid gland, which in some way brings about a hormonal change which allows the blenny to exist for some time in the open air.

▼ Porcupine Fish

This is yet another bony fish whose resemblance to a terrestrial animal has given it its colourful and highly descriptive name. Normally, these fish have an ordinary appearance, but when they are distributed or frightened, they draw in large quantities of water, swelling their bodies and erecting their spines, presenting a formidable challenge to a predator.

◄ **Goby**

Gobies are among the smallest of the bony fish. There are more than 500 species of goby, distributed all over the world, mostly in marine waters, although a few species are found only in brackish areas and estuaries. Their most notable distinguishing characteristics are their flattened heads with large, high-perched eyes and shortened snouts, and their forward displaced pelvic fins which are joined together to form a sort of sucker.

Gobies are bottom dwellers, seeking out hiding places for homes under rocks or debris. Some dig small burrows and others hide out among the corals. A few inhabit beds of seaweed and a handful of species live in shoals in shallow open water. They feed on small invertebrates, especially crustaceans, and will also eat any carrion they find. At the time of breeding, male gobies become highly territorial, defending the area which will become the nest in which the female will lay her eggs. Fertilization is internal, and the female, guarded by the male, lays her eggs in a clump, in which each one is fastened by a short stalk to a rock or other suitable surface. The female then leaves, and the male remains to guard the nest, driving off any intruders or possible predators, until the young hatch and disperse.

▶ **Cardinal Fish** *Apogon nemoptera*

The name cardinal fish includes a large number of small, related species, most of whom are found on coral reefs, mostly in the Pacific. The group takes its name from the earliest discovered members, which are bright red, but many members of the group, such as this pyjama fish, have only small amounts of red or none at all.

Cardinal fish readily assume associations with a variety of sedentary animals. Conches, corals and sea urchins are favourites, presumably offering protection, shelter and in some cases food, when the cardinal fish is able to feed on the bits and pieces thrown off by its associate.

Cardinal fish are listed among a large group of fishes which have in common the behavioural characteristics of so-called mouth breeding. In such species, one parent, usually the male, will keep the fertilized eggs in its mouth until they hatch. In some species, this behaviour occurs only when danger threatens, in others they are retained all the time. Both parents may share the duties in species in which the egg-holding is done only intermittently. The number of eggs held varies with size. In one Mediterranean species, up to 22,000 eggs may be held by the male. In some species of cardinal fish, newly-hatched young may return to the parent's mouth from time to time for shelter.

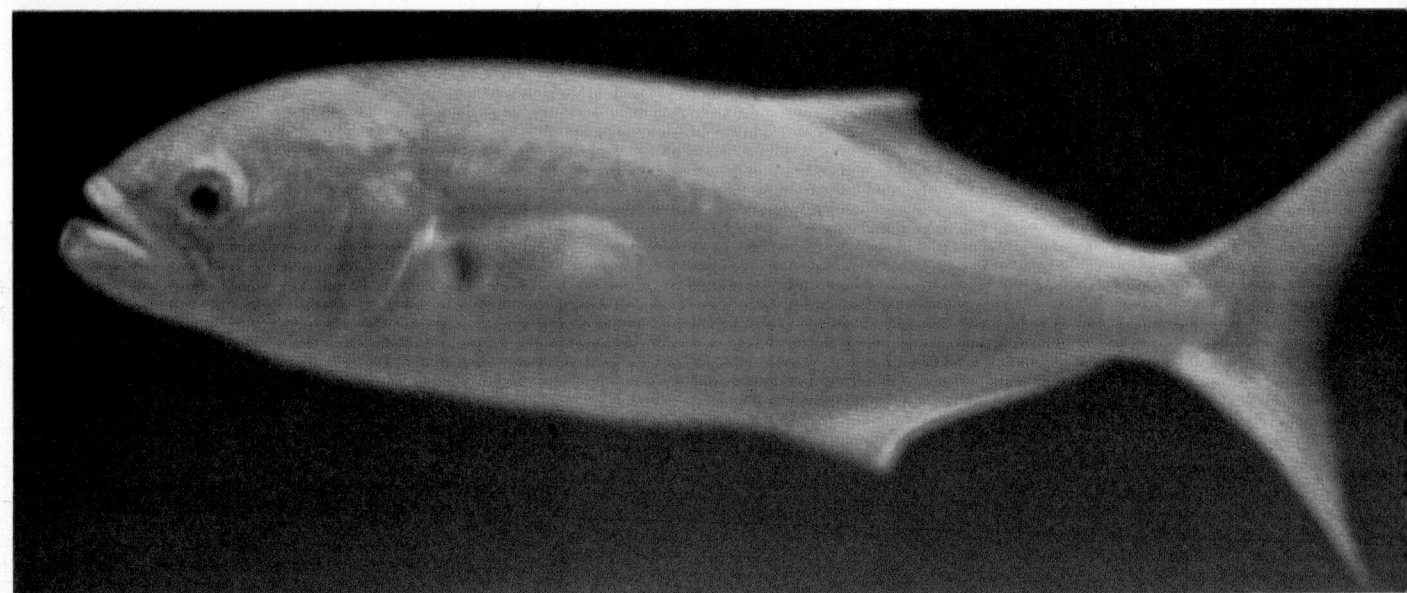

◄ Wrasse

Although brilliantly coloured, the wrasses are somewhat unique in that many of the 600 species are not tropical, but are found in colder, temperate seas, such as the coastal waters of Europe and North America. Above is an Atlantic bluehead, Thalassoma ruppelli, and below is a cuckoo wrasse, Labrus mixtus. One interesting aspect of this coloration is that it varies with sex and with the age of the individual, and can apparently vary from time to time, for example changing colour and pattern during courtship and mating. Wrasses are for the most part solitary fish rarely forming schools except for spawning aggregations in a few species. Like most solitary animals, they are highly territorial and very aggressive and will attack the fins and eyes of intruders. The male, or sometimes the pair, will build a nest of tangled weeds and other materials, or dig a shallow trough in the sand in which to lay the eggs, but, in spite of their territorial behaviour, wrasses do not appear to guard the eggs, but desert the nest as soon as the eggs are laid.

► Sole

The sole is another marine flatfish which, like the flounder, has come to lie on its left side, the mouth, fins and eyes adapting their positions and functions accordingly. The sole is best-known as a food fish and some people think that it makes the best eating of all fish. The sole is a bottom dweller of shallow seas with sandy or muddy bottoms. It prefers warm water, but some species are found either in northern or deep cold waters. The sole is a nocturnal animal. By day it lies in a trough on the bottom, which it digs by flapping its body up and down in a wave-like motion, at the same time stirring up mud and sand which settle on top of the animal and give it additional camouflage. It feeds entirely on bottom-dwelling animals – worms, crustaceans and molluscs – but is incapable of pursuing its prey and eats whatever it happens to encounter, rather than actively searching it out.

◄ Bluefish *Pomatonus saltatrix*

The bluefish is a medium-sized marine teleost, which lives in schools in tropical and sub-tropical waters. It is one of the most ferocious fish, attacking almost anything. It is therefore a popular game fish as well as an important part of many commercial fisheries. It has been extending its range in recent years, and has invaded the Mediterranean since 1945.

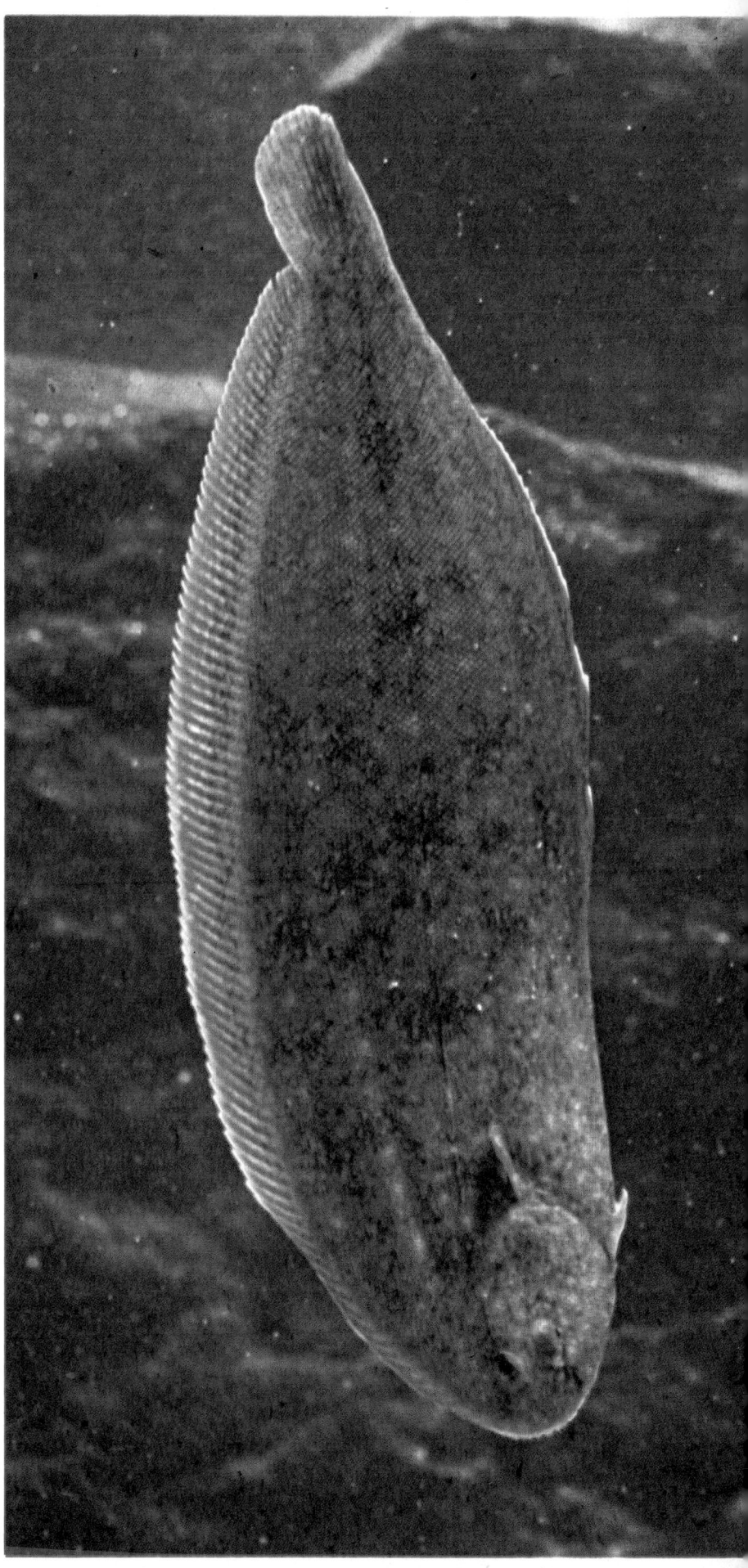

◄▲ Snapper *Lutjanus kasmira*
*Snappers are found in warm seas in all parts of the world.
There are more than 250 species. Some are important
food fishes, both for commercial fishermen and
sportsmen. They tend to move about in small shoals of
about 12, as is this group of Indian ocean snappers,
searching for food on a coral reef. They are night
feeders on the whole, actively pursuing living prey,
although it appears that they will eat almost anything.
They are known to eat representatives of almost every
class of marine invertebrate, as well as large numbers of
smaller fish, underwater plants and organic wastes. The
bulk of the diet seems to vary with what is most abundant,
rather than having a particular type of preferred food.
Little is known about the snapper's breeding habits ;
different species seem to have different breeding patterns.
Some breed only once a year, others twice, while yet
others seem to be able to breed throughout the year. In
some regions, snappers have a reputation for being
poisonous and therefore inedible, but this is probably
due to their occasional habit of eating a particular alga
which can cause cramps, nausea and even temporary
paralysis, or by their eating an herbivorous fish which has
fed on the alga, rather than it being a function of some
poison produced by the snapper itself.*

◄▼ Sole
*This is a photograph of a metamorphosing sole, about
sixteen times life size, in which the migration of the left
eye up to the right side of the body is not yet quite
complete.*

▶ Angelfish *Pomacanthus paru*
*Angelfish are relatively well-known, but as this name is
commonly applied to three different groups of fish, it is
important to distinguish just which group is being
discussed. One, the most commonly encountered, is
freshwater, and is a favourite with aquarists. Another is a
relative of the sharks, and therefore an elasmobranch,
and is sometimes called the mouthfish as well as the
angelfish. The third is a group of marine angelfish. The
marine angelfish, and their near relatives the butterfly
fish, together number about 150 species. They live mainly
in warm shallow seas, singly or in pairs, around reefs,
rocks and corals. A few species enter estuarine waters.
Angelfish are most notable for their bright colours and
patterns. Colours and patterns may differ with age or
between the sexes. This photograph of a French
angelfish shows the kind of coloration as well as the rather
bizarre form of these fish. The small mouth is armed with
strong teeth, used for crushing the small invertebrates on
which angelfish feed. The reason for their bright colours
and complex patterns has not been fully explained. They
are not aggressive, dangerous or poisonous, so it seems
unlikely that this display is intended as a warning to
predators. On the contrary, it is reasonable to assume that
angelfish call attention to themselves in this way, and that
this method of finding a mate outweighs the disadvantages
of attracting predators.*

◄ Surfperch

The surfperch is very similar in most respects to its near relatives the freshwater perches. There are about 25 species which, for the most part, live in shallow seas, although some are occasionally found in tidal pools, others at depths of up to 100 fathoms and at least one species invades fresh water. Two species inhabit the coastal waters of Japan and Korea; the rest are found on the Pacific coast of North America, from California to Alaska. The breeding habits of the surfperch are quite extraordinary. The young males are sexually mature at birth, and mate with the immature females when only 2 days old. The sperms are retained by the female until she reaches maturity soon after. The sperms then fertilize the eggs while the latter are still in the ovary, and the fertilized eggs are released from the walls of the ovary and are retained within its cavity while they develop. Embryonic development is rapid. A gill opening soon develops, and cilia surrounding and lining it drive a current of liquid into the primitive digestive tube. In this way the embryo derives its nourishment and its oxygen. Later as the fins develop, they too absorb food from the ovarian fluid, and finally the young surfperch, now able to feed, eat some of the cells lining the ovarian cavity, as well as any remaining sperm. Thus the surfperch is unique among marine teleosts, in that it has internal fertilization, embryonic development in the ovarian cavity, gives birth to live young, of which the males are already sexually mature, and it practices delayed fertilization.

► Hatchet Fish

There are two types of hatchet fish, marine and fresh water. There are 15 species of marine hatchet fish, which are distant relatives of the salmon. They are all small and weigh only a few grams on average. This photograph does not show the reason for the name hatchet fish. The head is large, and is merged with the larger body; the tail is narrow, and seems not to belong to the fish. The overall impression in a sideways view is that of a hatchet. As might be assumed from their prominent eyes, hatchet fish live in the twilight zones of the ocean, from 50 to 260 fathoms in all warm and temperate seas. Their eyes are at least as sensitive as the human eye, and are probably extra-sensitive to the blue and green rays which penetrate to the depths at which they live. It has been suggested that the hatchet fish's eyes may even be telescopic, but there is no direct evidence for this. Like many animals living at great depths, the hatchet fish has light-producing organs. These are on its abdominal region, but the reason for their presence is not clear as the eyes are so high-set on the head that it is doubtful that they could see any light produced by that part of the body.

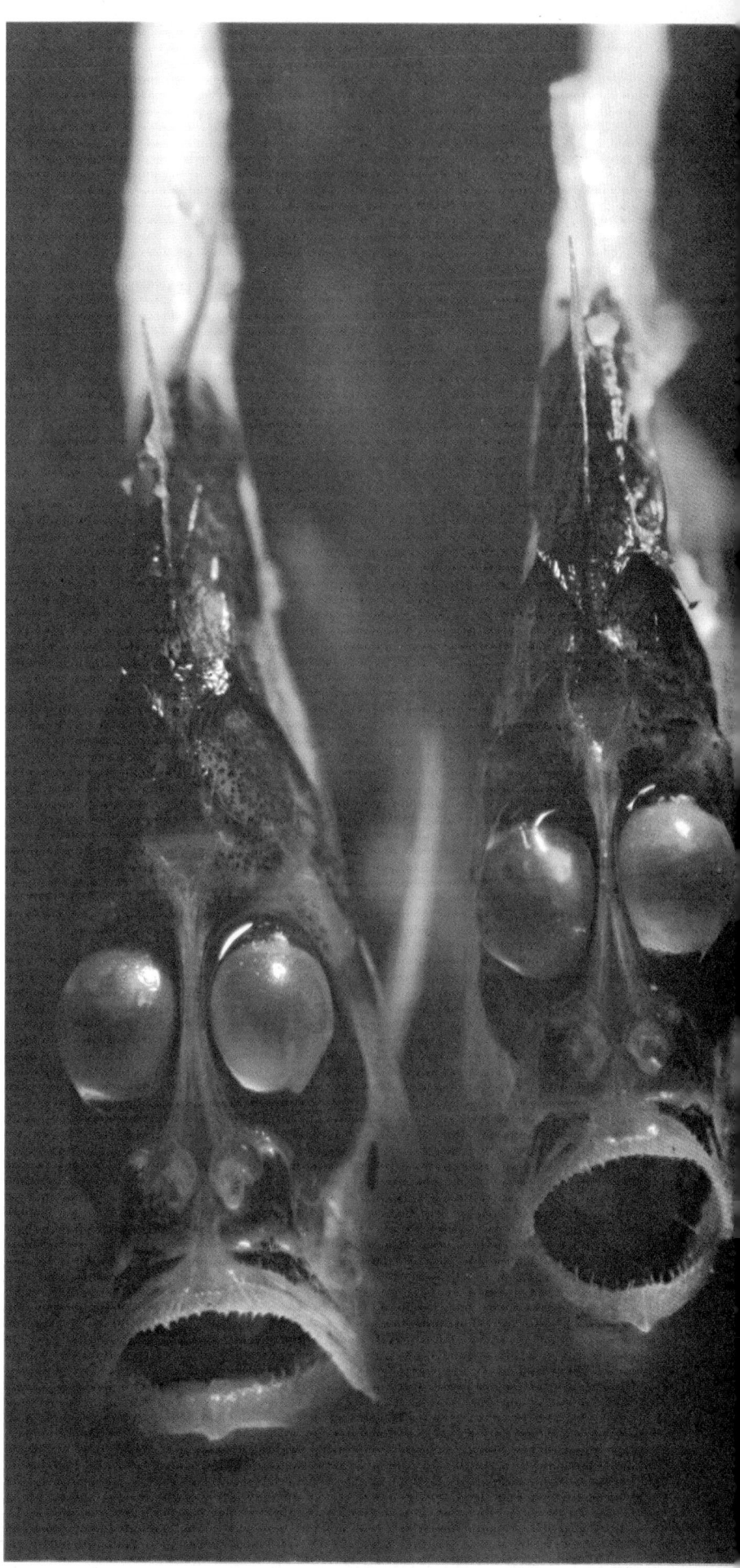

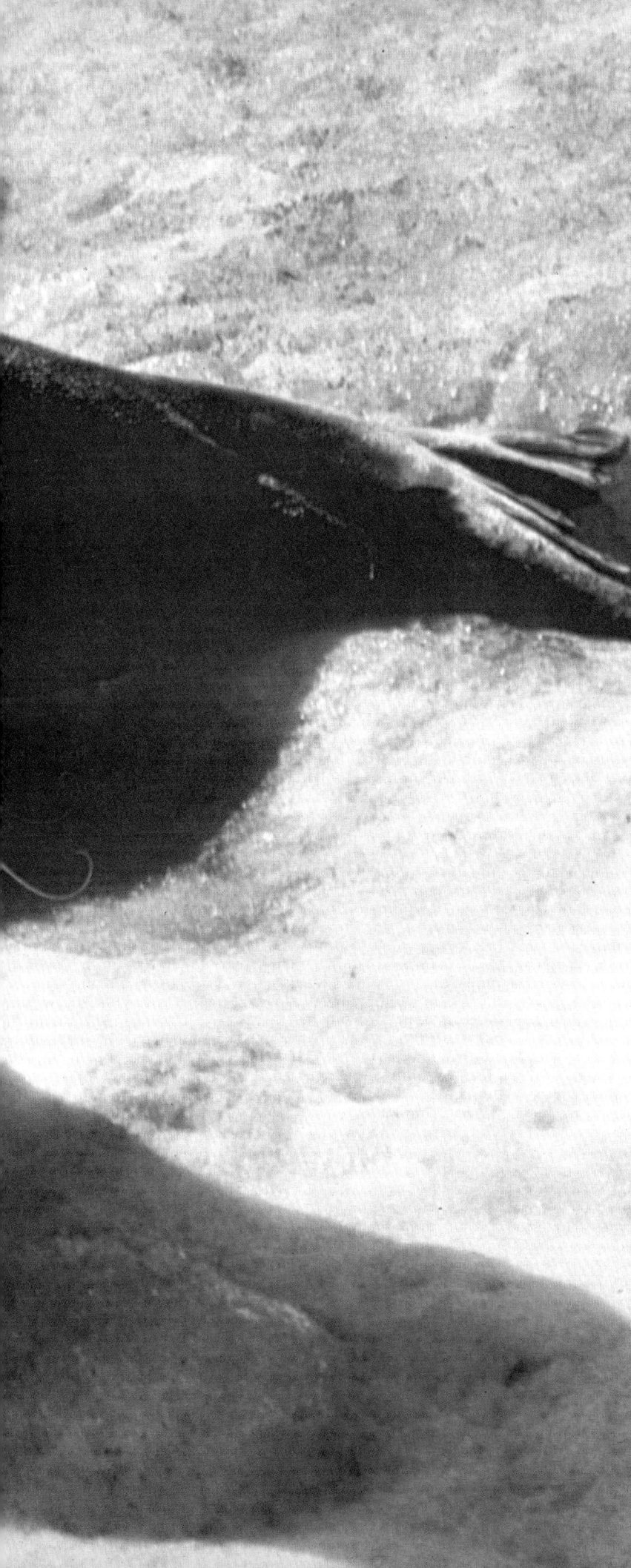

Birds and mammals

With retrospective knowledge about the evolution of the vertebrates and their colonization of the sea on two separate occasions, first by the elasmobranchs and then by the bony fish, it seems inevitable that the next step would be the conquest of the land. This was inevitable but it was a long, slow process which took millions of years. The transition from fresh water to land was a momentous step, opening up whole new areas for the vertebrates to exploit. It was also very difficult because conditions in the air are so different from those in the water. Air does not provide water and salts and its abundant oxygen must be extracted from it in a new way, and so major physiological and anatomical changes became necessary. Air does not have the constant temperature of water, so that an animal living on the land has to be able to cope with large changes in the ambient temperatures.

There is not the space to discuss fully the slow conquest of the land by the vertebrates. A small specialized group of bony fish gave rise to the amphibians, the first vertebrates to exploit the land. The reptiles which arose from them were the first group of vertebrates to free themselves entirely from the water. The amphibians were tied to the water for breeding, but the reptiles, with the development of an egg which was covered by a tough protective shell, were not. The reptiles however, had one major problem. They could not exert any particular control over their body temperatures, so that they and the amphibians were restricted to areas which were warm enough to support life. The birds and mammals, both of which arose from reptilian ancestors, solved this problem. They are able to maintain a high body temperature which is independent of the environmental temperature, and this has made it possible for them to exploit almost every known ecological niche.

This exploitation of all parts of the environment is known as adaptive radiation, and in the course of this process it was only a matter of time before some birds and mammals acquired specializations which allowed them to return to their ancestral aquatic home. The most aquatic birds, the penguins, have become flightless in the process of becoming aquatic. The vast range of mammals – seals, walruses, sea lions, whales and so on – will also be discussed. Their adaptations to aquatic life have been many. They have become fish-like in many ways, an example of convergent evolution. For the most part, the truly marine birds and mammals are confined to polar regions, where competition from other animals is minimal or absent.

Weddell Seal *Leptonychotes weddelli*
There are four species of Antarctic seal, of which the Weddell seal is the best known. Part of the reason for this obvious from this photograph. They are not afraid of man and are easy to approach. Also, they breed near the coastline and live under the relatively transparent sea ice, so they have been quite easy to study.

275

▶ **Sea Lion** *Eumetopias jubatus*
There are five species of sea lion, most of which are found in the region of the Pacific Ocean. Their breeding grounds are along the western coast of North America, the northern coasts of Alaska, Japan, China and Russia, the southern coast of Australia, the Galapagos Islands and the west and east coasts of South America. This is a photograph of a rookery of the Steller's sea lion, along the North American coast.

▼ **King Penguin**
The king penguin is another Antarctic species, and is often confused with the emperor penguin, a near relative. The king penguin is the smaller of the two, and is distinguished from the emperor penguin by the size and shape of the yellow patches on the side of the head. The king penguins have a wider range than the emperors, living as far north as the ice-free seas around the Falkland Islands. They live at sea when not breeding, and can swim long distances.

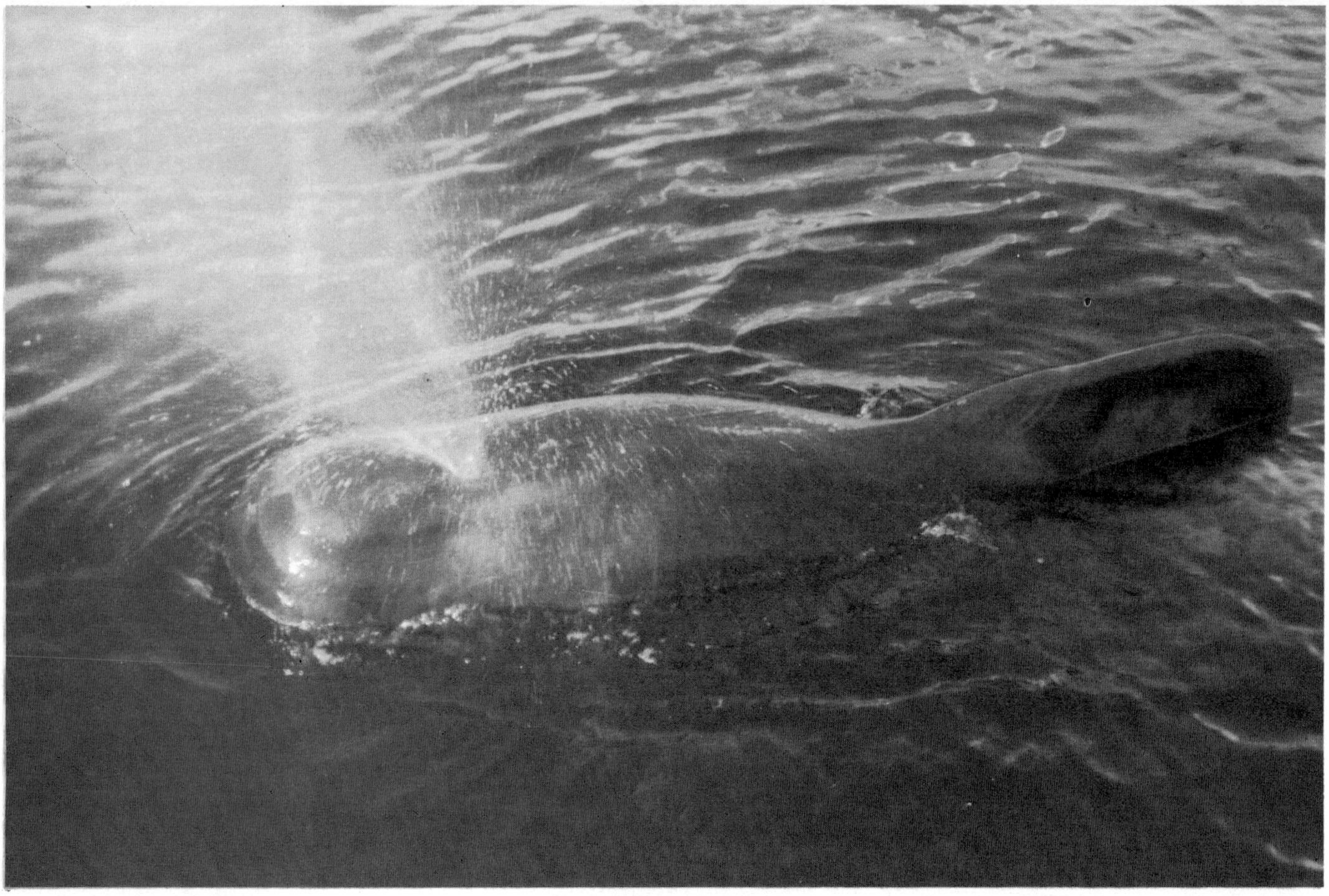

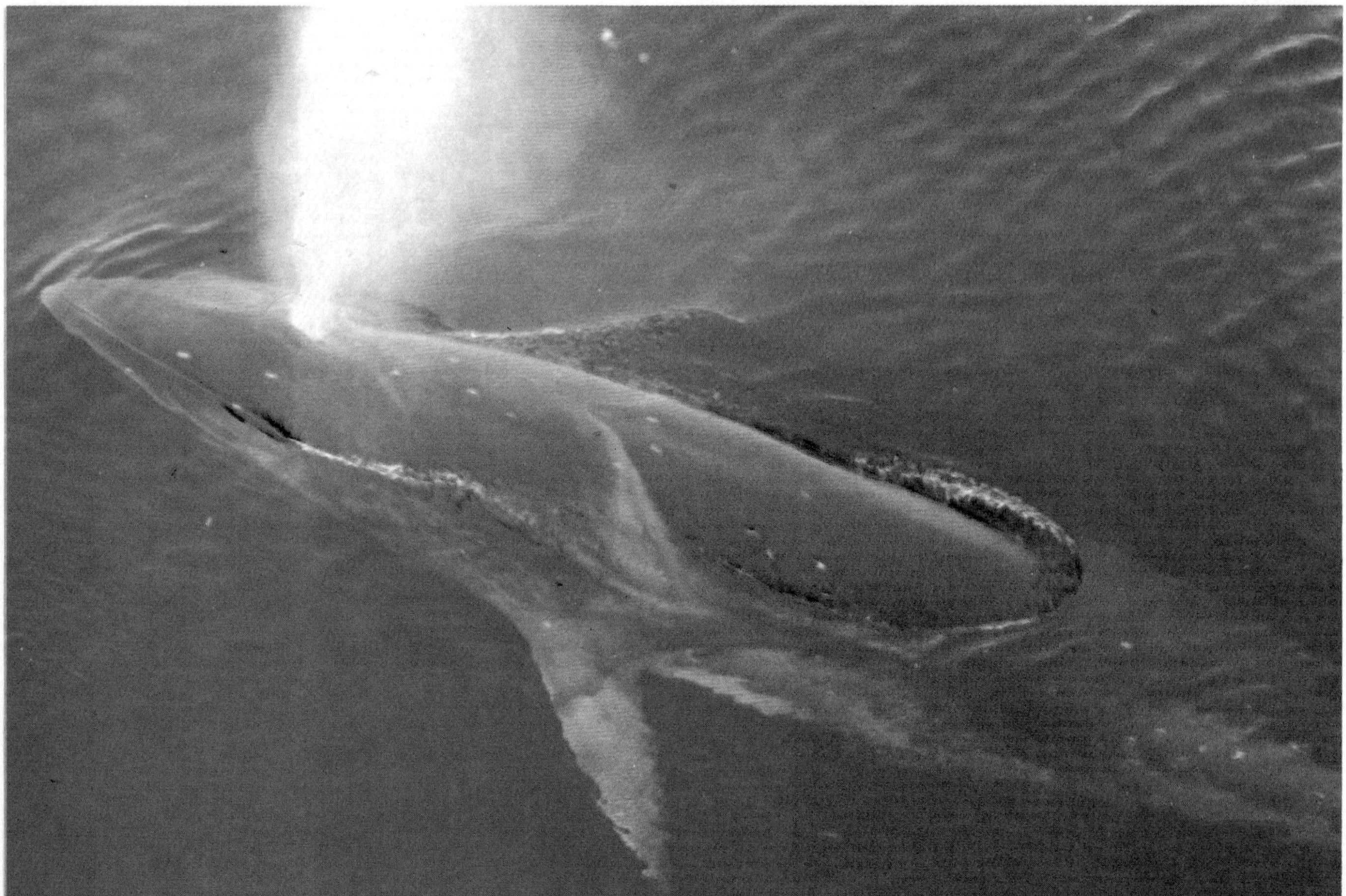

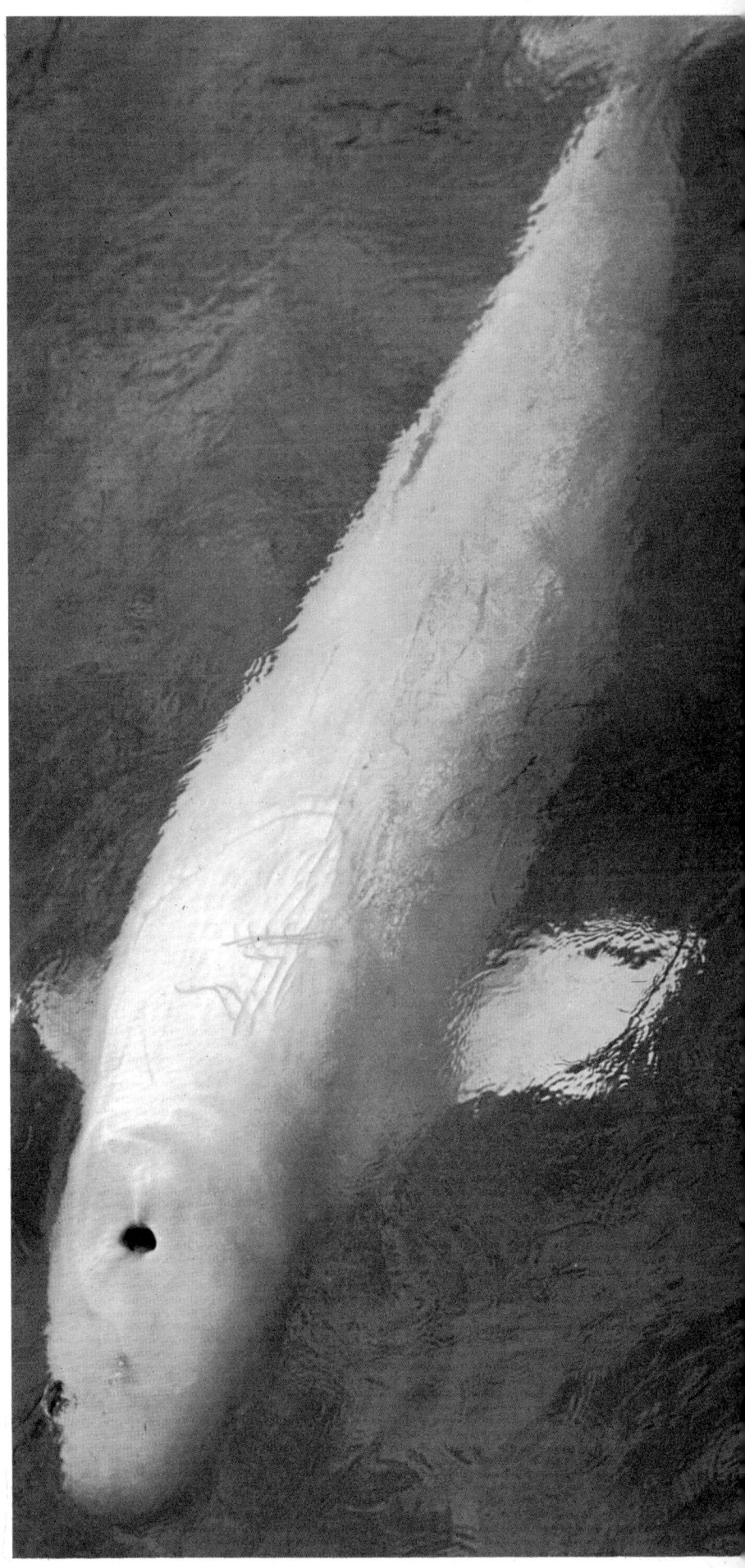

◄▲ Pilot Whale

The pilot whale is a large dolphin which may reach a length of up to 10m. There are several species of pilot whales, which are black except for a white patch under the chin. They live in large schools which may number hundreds or even thousands of individuals. The name comes from their follow-the-leader habit, and it has been suggested that a single dominant male may lead each pack, but it seems more likely that the pack will follow whichever animal is in front at any time. Pilot whales are armed with strong teeth which they use for securing their prey. They eat mainly squid and cuttlefish, although some bony fish probably make up a part of their diet. Schools of pilot whales migrate annually, moving north in the summer after new food sources, and south in the winter as the water becomes colder. They are found in all parts of the world, including the Mediterranean, but are absent from the cold polar seas.

◄▼ Rorqual Whale *Balaenoptera acutirostrata*

There are four species of rorquals or fin-backed whales. They are close relatives of the blue and humpback whales, and the three types are generally known as baleen or rorqual whales. This is a photograph of one of the rorquals, the mink or lesser rorqual whale. Rorquals live in schools of up to several hundred animals, although groups like this are rarely seen nowadays, since the baleen whales have been the most severely depleted by over-fishing. Although they are common in the polar regions, most rorquals migrate to warmer waters in the winter, congregating in areas where food is plentiful, returning to the colder regions in the summer, although some individuals seem to stay in warmer waters throughout the year. Rorquals feed in two ways. Those which gulp their food do so by taking a mouthful of water, shutting the mouth and expelling the water through the baleen. Other species sieve their food by rushing through the water with their mouths open, the water passing rapidly through the baleen and the food catching in it. When they have a mouthful, they shut their mouths and swallow.

► Beluga *Delphinapterus leucas*

The beluga, along with the narwhal, forms one of the sub-divisions of the dolphin and porpoise family. They are the most easily identified of all the cetaceans, because they are the only all-white members of the group. Until about four years of age, they are grey or mottled white, but soon lose this coloration. They are natives of the seas around the North Pole, although in severe winters they move south, and have been seen in Scotland, Ireland and Japan. Belugas are toothed whales, and feed on shrimps, cuttlefish, crabs and a variety of bony fish. They were once numerous, with herds of up to 10,000 animals being reported, but they too have suffered from overfishing and herds now number about 12 or so. They have few natural enemies, other than man, the principal one being the killer whale, who will attack a full-grown beluga, although it prefers to take on a smaller one if possible. Belugas are said to panic at the sight of a killer whale.

Bottlenose Dolphin *Tursiops truncatus*

The bottlenose dolphin, or common porpoise as it is known in North America, is one of the best known whales. It is the most common whale found in the North Atlantic, from the Caribbean and African coast as far north as Scandinavia. It is also common in the Mediterranean. They live in schools or packs of mixed sexes and ages, and, although there does not seem to be a leader of the pack, the males, at least, observe some sort of hierarchical organization in which some males are dominant over others. They are also noted for being co-operative animals, assisting injured members of the pack by holding them at the surface and allowing them to breathe, and providing the same sort of support for females at the moment of birth. This sociability is evident even in times of play. The jaws of the bottlenose dolphin are well armed with small teeth which help it secure and break up its food, which consists mainly of cuttlefish, bony fish and shrimps.

Like all cetaceans, dolphins are relatively slow breeders. They do not become sexually mature until five or six years of age, and the gestation period is about one year. The young calf sticks close to the mother for some time, both for feeding and protection. For the latter, the calf may also move towards an 'aunt', another female which the calf can recognize and which is the only adult that the mother will allow near her young. The young dolphin does not begin to take solid food until at least five months old, and may remain with the mother and not be fully weaned until at least 18 months of age.

Dolphins, like most other cetaceans, are powerful swimmers and jumpers. The tail is the main organ of locomotion and, unlike that of fish, moves in a vertical rather than horizontal plane. The forelimbs are used mainly for balancing. The hind limbs are of minimal use to whales.

They have a poor sense of small but very acute hearing. It was once thought that their eyesight was no use out of water, but this now seems untrue.

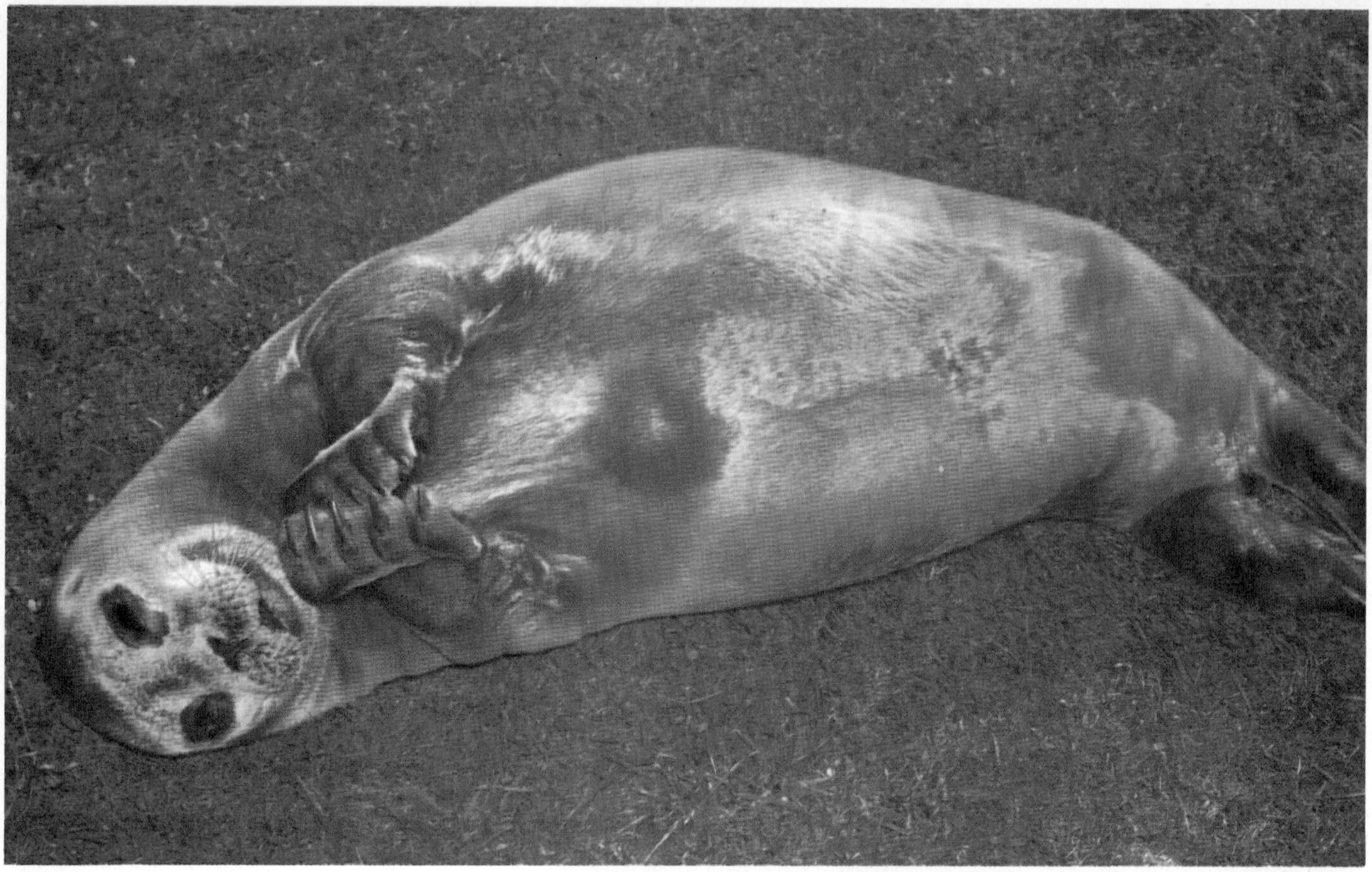

▲ Common Seal *Phoca vitulina*
*The common seal is found along the coastlines of
temperate and cold seas in the northern hemisphere.
They prefer any shoreline where they can easily haul
themselves out of the water. This is why they are
particularly found in inlets and estuaries, although some
herds are found on rocky, exposed costs. This photograph
gives some idea of why the seal is an ungainly animal on
land. The hind flippers, completely adapted to aquatic
life, cannot be tucked up under the body to assist
locomotion on land, so that the seal is reduced to pulling
itself along with its front flippers. Seals spend the winter
at sea away from the mainland, but return to sheltered
areas along the coast. The pups are born quickly, at
low tide, and must be able to swim immediately. They
take to the water – some may even be born in the water
if conditions are adverse. They spend the first few days
learning to swim and dive, rarely, if ever, coming ashore.*

Walrus *Odobenus rosmarus*
*There are two subspecies of walrus, the Atlantic walrus and the Pacific. The Pacific
ones are the larger. Walruses are found in the northern hemisphere, forming a rough
ring around the Arctic Circle. Herds which live farthest north migrate south in the
winter, returning to their northern grounds to breed in the summer. Walruses are social
animals, as these two photographs show. Large herds, composed of families of 100
animals or more, are found, like the one on the right. The herds are made up of cows,
calves and young bulls. Adult males form separate herds and come into contact with
other walruses only in the breeding season. Large herds were once common, but
persecution by man has reduced their numbers, and the remaining walrus herds are
tending to avoid land and keep to the ice floes. Strict conservation has saved the
Pacific walrus but the Atlantic subspecies is still in danger.*

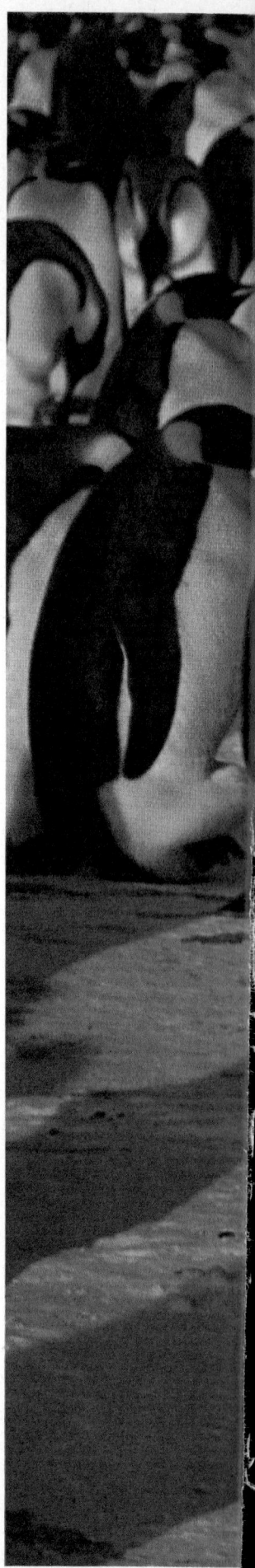

Adélie Penguin *Pygoscelis adeliae*

*The Adélie penguin, along with the emperor penguin, is
found only in the Antarctic. Some other species of
penguins are also found there, but they are occasional
visitors who spend most of their time and have their
breeding grounds further north, some species being
found as far north as the Equator. Penguins are birds of
the southern hemisphere, however, and are not found in
the Arctic. All penguins are flightless birds, more
comical than clumsy on land. They are superbly adapted,
physically and behaviourally, to their sub-zero habitat.
Penguins spend much of their time at sea. They are
excellent swimmers, the wings having become adapted as
flippers, the body shaped hydrodynamically. Feather
structure is adapted to this aquatic life and beneath the
skin is a thick layer of blubber, which not only insulates
the penguin, but provides a source of food during periods
when the bird is not feeding. The penguin is a gregarious
bird, coming ashore in October, at the beginning of the
breeding season to form rookeries, large colonies of
nesting penguins.*

*In the vast and desolate wastes of Antarctica, there is
no food to be found on the land mass, so that all
inhabitants of this region must do their feeding at sea. At
the point where the warm waters of the north meet the
cold waters of the Antarctic, there is a region of abundant
plant and animal life, and it is this which provides the
diet of the Adélie penguin. It is composed primarily of
small molluscs and crustaceans, but most other aquatic
animals will be eaten if they are available.*

*They are commonly seen tobogganing across the ice,
lying on their fronts and propelling and steering with their
flippers and their feet.*

▶ **Emperor Penguin** *Aptenodytes forsteri*

*The emperor penguin, as its name implies, is the largest
of all the species of penguins. Their breeding grounds are
in Antarctica, although in the summer they move north
to feed, and have occasionally been found as far from
home as the tip of South America. The adults begin
breeding in the winter. Emperor penguins do not build
nests, so that there is no territorial behaviour. Each
female lays one egg which is incubated, apparently
without great enthusiasm, by both parents who takes
turns to carry it on their feet where it is covered by a fold*

*of skin. Soon the female leaves and goes to sea, and the
male is left to incubate the egg on his own for two months
of the bitterly cold winter, the colony huddling together
for warmth as the temperature drops. The females return
to feed the chicks, the males then leaving to feed.*

*Penguins have no terrestrial enemies as there are no
predatory land animals in the Antarctic, but weak young
chicks fall prey to great skuas, and juveniles and weak
adults to leopard seals in the water. The emperor
penguin was once thought to be in danger of extinction
but is now known to be flourishing.*

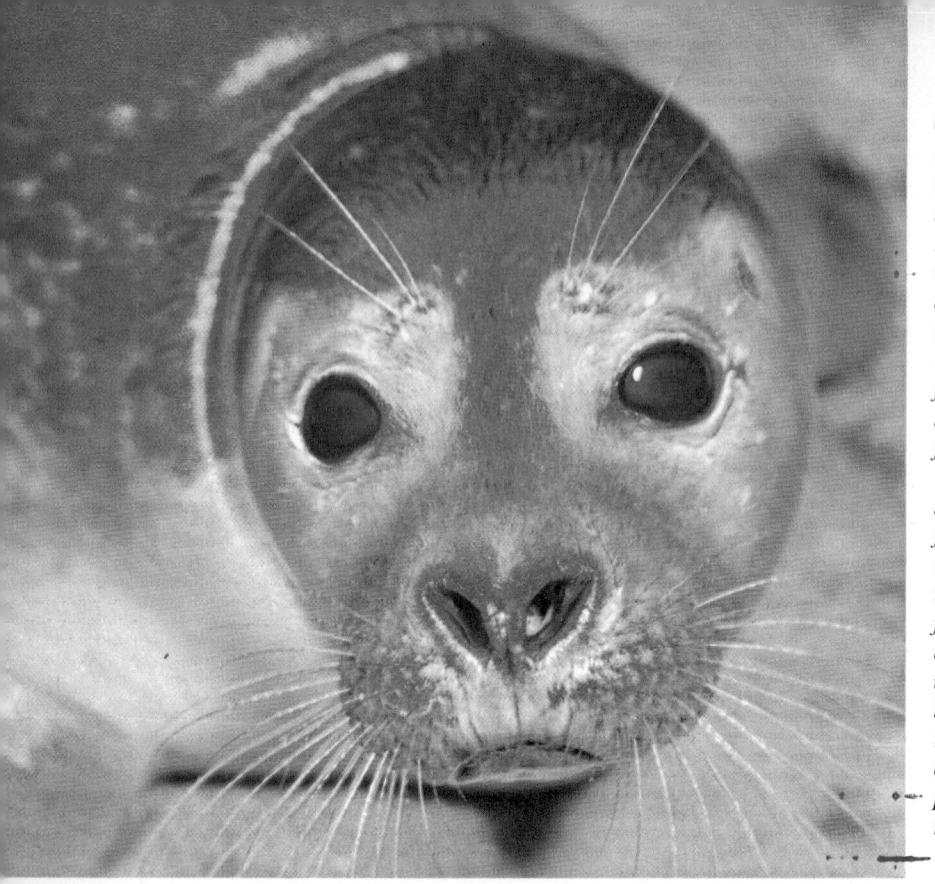

Common Seal

Seals, like other aquatic mammals, are well-adapted to their almost entirely aquatic life. Their streamlined bodies reduce resistance for movement through the water. Even the external evidence of the ears has been lost, as these pictures show, to make the head as hydrodynamic as possible. The external ear is reduced to a slit behind the eye. The facial whiskers are extremely sensitive to vibrations. This assists the seal when it is hunting for food. Specially developed muscles in the snout allow the seal to close its nostrils when it dives, to prevent water from entering the lungs.

Like all other seals, the primary diet of the common seal is fish. Newly-weaned seals begin their diet of solid food with shrimps and other small crustaceans, and as they get older they add molluscs, crabs and fish. Most of their adult diet is made up of food found on the sea-bed – flatfish, gobies, soles, and the like, but seals also pursue and catch squid, mackerel, salmon and trout. It is this unfortunate taste for fish of high commercial value that brings seals into conflict with fishermen. Some seals turn to net robbing, inflicting a lot of damage on nets as well as on the fish in them. However, this damage does not prove the case for the wholesale slaughter of seals recommended by some.

King Penguin

Like the emperor penguin, the king penguin (bottom and right) has breeding problems, because of its size and the length of time it takes the chicks to become independent. Rather than breeding during the winter, like the emperor, the king penguin has solved this problem by breeding further north where the sea does not freeze and the adults have more freedom. The eggs are laid in the spring, after the males have indulged in a lot of display and courtship rituals which attract a suitable female. The nest is built but the male, who is left to incubate when the female returns to the sea to feed after laying, does not move about with the egg. He defends a small area for about two weeks, until the female returns and relieves the male. Thereafter the parents take turns incubating and feeding until the chick hatches about eight weeks after laying. Both parents share the feeding and protection of the young throughout the following winter. At first the young grow rapidly, being fed plenty of fish at each feeding, but as food becomes scarcer in the course of the winter, they lose weight. With the coming of spring and an increase in the food supply, there is another growth spurt, and the young moult into their adult plumage. Two months later when food is abundant, they move to the sea and quickly learn to swim and fish for themselves. At first they spend most of their time at sea, but as they get older they spend more time ashore, practising their courtship displays in preparation for their first breeding season. Like many large sea birds, king penguins are slow breeders. They do not become sexually mature until they are six months old, and because of the breeding schedule, and the long post-hatching period during which the adults look after their young, there is only one chick per pair every two years.

◀ **Adélie Penguin**

*The breeding behaviour of the Adélie penguin, although
similar in some respects to that of the emperor and king
penguins, has its own distinctive aspects. The adults
spend most of their time in the water, but come to land
on the Antarctic continent in September or October.
They have spent the winter feeding, and are in good
breeding condition at this time. They make their way over
land to the breeding grounds or rookeries which are
usually located on rocky headlands. The penguins return
regularly to the same breeding grounds, and on arrival
each male searches out his old nest. Young males take
over unused ones. The males occupy their nests, defending
them from intruders and displaying at the same time.
The females return to their old mates; new females,
or ones whose mates have not returned, are attracted
to other males by the vigorous display and defence
routine. When the pair is formed, the female remains
at the nest while the male collects pebbles from the beach
for repairs and renovations. He takes the pebbles to the
female, who builds a ring around herself.*

*Two eggs are laid. As soon as they are laid, the female,
who has not fed for two or three weeks, returns to the
sea while the male incubates the eggs and defends the
nest. Within a couple of weeks she returns and relieves
the male, who has not eaten for as many as six weeks,
and has lost half his weight by this time. He goes off to
feed for a week or more, returning to relieve the female.
The parents in turn incubate the eggs until they hatch,
and then take turns going to sea to collect food for the
chicks. For the first few days the chicks stay under the
guarding parent, but they grow quickly and soon come
out to stand by the nest. When they are about a month
old, they leave the nest and gather in a crèche, a sort of
large public nursery for young penguins. They stay in the
crèche for some time, while the parents continue to feed
them. For feeding, the parent collects a chick and leads it
away from the crèche. The chick is introduced to the
outside world by leading it on little forays which adjust it
for the time when it will leave the crèche and its parents,
go to sea and begin to fend for itself.*

▶ **Walrus**

*This photograph of a mature male walrus shows well two
of its characteristics. Body hair is almost completely
absent, the hide being toughened and deeply folded.
The only hairs left are the coarse vibrissae of the snout,
whiskers which the walrus uses for detecting shellfish, a
principal food. The large tusks are multi-purpose. They
are used as offensive and defensive weapons and for
digging food out of the mud. They are used for breaking
up ice to make breathing holes. The walrus also uses them
for locomotion, digging them into the ice to pull himself
along, or using them like mountain climbers use hooks,
driving them into the edge of the ice and then pulling
himself out of the water.*

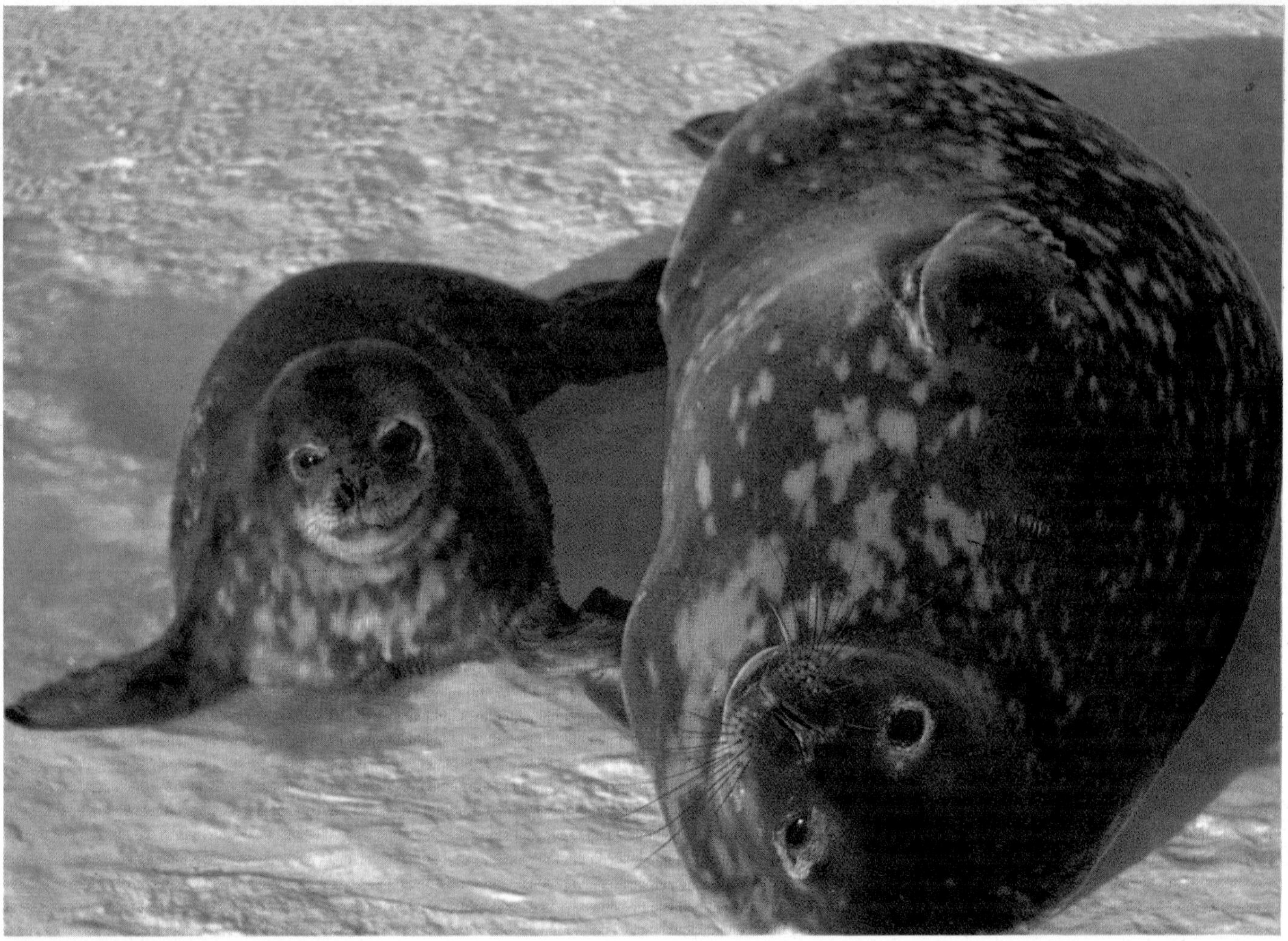

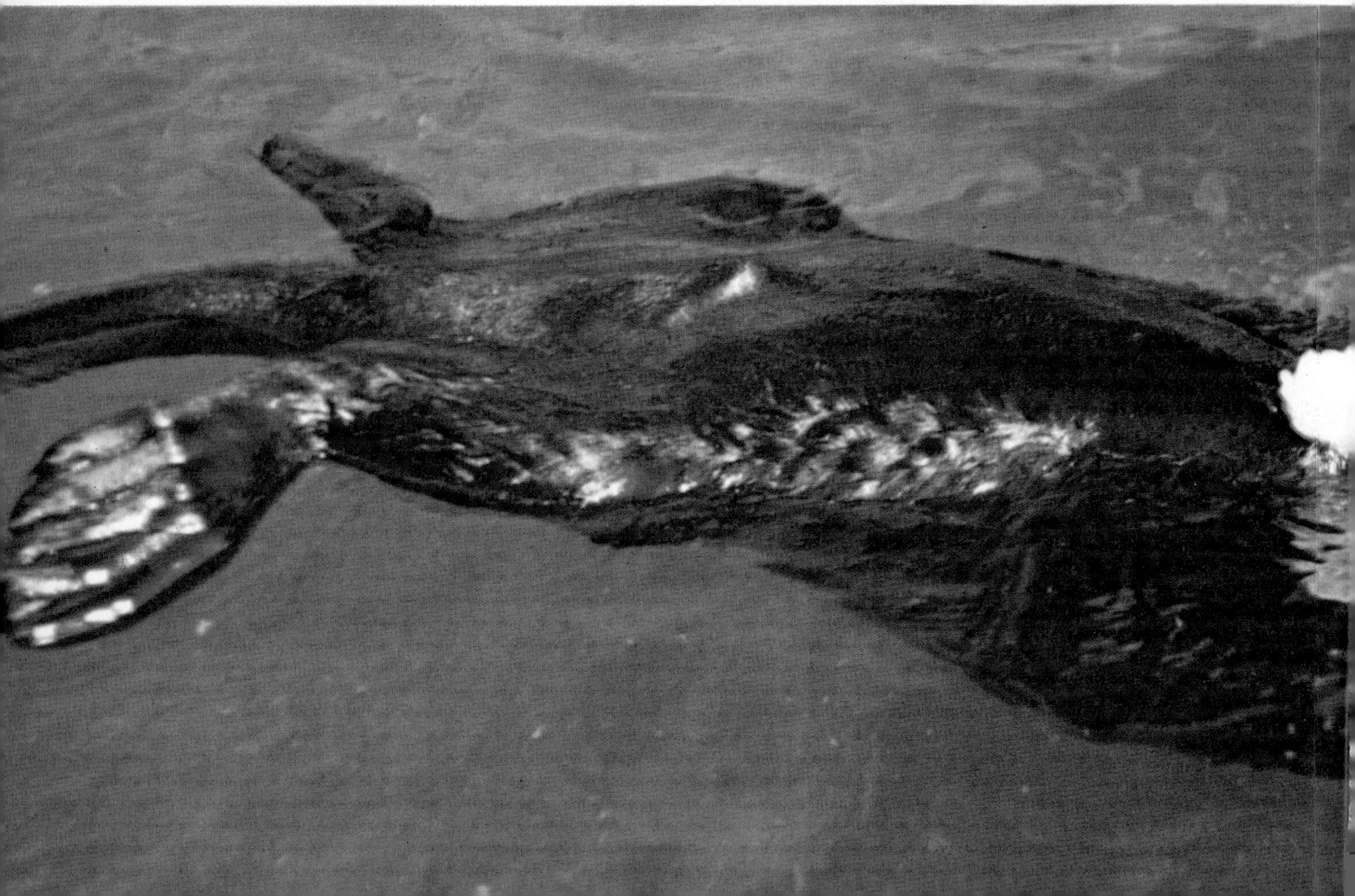

◄◄ **Weddell Seal**

The Weddell seal is the mammal found further south than any others. Although they are usually found within sight of land they spend most of their time in the water, which is usually warmer than the air.

During the winter, most of their water range is frozen to a depth of half a fathom or more, and the seals must be able to find breathing holes. Natural holes are kept open and, if necessary, new ones are made by chewing through the ice with the teeth in a sort of sawing motion. This is of course very hard on the seal's teeth, and so tooth wear and damage is probably the principal cause of natural mortality.

Young seals are born in the spring. The females haul themselves up on to the ice, or sometimes the land, and the single pups are born a few days later. The pup has no insulation layer of blubber and is born directly on to the ice, going through a massive temperature drop at the moment of birth. The pup is not weaned for almost two months, and during this period the mother does not feed, but loses weight, while the pup puts on weight at almost the same rate. The male plays no part in the rearing of the pups.

Sea Otter *Ennydra lutris*
The sea otter, which is one of the largest otters, is found along the North American and Asian coasts of the Pacific. It is exclusively marine, and spends most of its time at sea, usually in small herds, within sight of land (above).

The posture shown by this feeding female sea otter (below) is a typical one. Crustaceans and molluscs, with the occasional fish, are the favourite foods, and the otter will dive down to 16 fathoms when hunting. Once the prey is captured, the otter returns to the surface and rolls over on its back, holding the food on its chest and eating it at leisure.

►► **Harp seal** *Pagophilus groenlandicus*
As its scientific name implies, the harp seal, sometimes called the saddleback or the Greenland seal, is found in the northern hemisphere, in the Arctic and North Atlantic Oceans. The large populations of Labrador and Newfoundland have been the most intensively studied. Early in the year adult seals move south as far as the Gulf of St Lawrence, haul out on to the pack ice, and there give birth to a single pup in late February or early March. The males join the females after the pups are born, and the juvenile, non-breeding seals arrive from the north soon after the males. The pups are born with a pure white coat which they keep for about a month before moulting into a grey coat speckled with dark grey and black markings. Both these coats have a high commercial value, and because of this, conflict has arisen between conservationists and sealers. Up to 300,000 pelts are taken each year. The numbers have been severely reduced so that recent controls now limit the numbers that can be taken and the areas in which sealing is permitted. As soon as the pups are weaned, they are abandoned by their parents and the entire population moves north again to spend the summer feeding.

Common Names Index